Wireless Location in CDMA Cellular Radio Systems

THE KLUWER INTERNATIONAL SERIES
IN ENGINEERING AND COMPUTER SCIENCE

WIRELESS LOCATION IN CDMA CELLULAR RADIO SYSTEMS

JAMES J. CAFFERY, JR.
University of Cincinnati
Cincinnati, OH, USA

Foreword by Gordon L. Stüber

Springer Science+Business Media, LLC

Library of Congress Cataloging-in-Publication Data

Caffery, James J.
Wireless location in CDMA cellular radio systems / James J. Caffery, Jr.
p. cm. -- (The Kluwer international series in engineering and computer science ; SECS 535)
Includes bibliographical references and index.
ISBN 978-1-4757-8425-1 ISBN 978-0-306-47329-6 (eBook)
DOI 10.1007/978-0-306-47329-6
1. Code division multiple access. 2. Cellular telephone systems. 3. Mobile communication systems. 4. Automatic tracking. I. Title. II. Series.

TK5103.452.C34 1999
621.3845--dc21 99-048309

This printing is a digital duplication of the original edition.

Originally published by Kluwer Academic Publishers in 2002
Softcover reprint of the hardcover 1st edition 2002

Second Printing 2002.

Printed on acid-free paper.

Contents

Foreword

With the increasing market penetration of cellular telephones, the number of E-911 calls placed by cellular telephones has grown considerably. This growth in E-911 calls led to a 1996 FCC ruling requiring that all cellular, PCS, and SMR licensees provide location information for the support of E-911 safety services. The provision of such location information is to be implemented in two phases. Phase I, whose deadline has already been passed, requires that wireless carriers relay the caller's telephone number along with location of the cell site and/or sector serving the call, to a designated Public Safety Answering Point (PSAP). This information allows the PSAP to return the call if disconnected. Phase II, to be completed by October 1, 2001, is much more stringent and requires that the location of an E-911 caller be determined and reported with an *rms* location accuracy of 125 m in 67% of the cases.

The applications of wireless location technology extend well beyond E-911 services. Location information can be used by cellular telephone operators themselves for more effective management of their radio resources, so as to achieve greater spectral efficiencies. Resource management algorithms such as hand-offs between cell sites, channel assignments, and others can all benefit from subscriber location information. Location information obtained from vehicular based cellular telephones can be used as an input to Intelligent Transportation Systems (ITS), and in particular traffic management and traveler information systems. Law enforcement agencies can also benefit from accurate location information by discouraging the use of cellular phones for criminal related activities.

This book provides a comprehensive treatment of the technical challenges in wireless location technologies. It is an outgrowth of a research project at the Georgia Institute of Technology funded by GTE in the early 1990s. We are indebted to Drs. Art Giordano and Khaled Dessouky of GTE Laboratories for presenting us with the original goal of establishing cellular radio as a medium of choice for ITS applications. It soon became apparent to us that wireless location would be an essential ingredient of cellular-based ITS applications. Our research then focused on time-based methods for radiolocation in IS-95 CDMA cellular systems. Later the 1996 FCC ruling brought heightened significance to our radiolocation research. All of this work ultimately lead to Jim Caffery's Ph.D. dissertation supplemented by his subsequent work at the University of Cincinnati.

I believe that research into the difficult problem of wireless location will continue for some time. Location systems must be developed for many different wireless standards, and all good location systems must combat the effects of multipath, non-line-of-sight propagation, and multiuser or co-channel interference.

Gordon L. Stüber
Professor
Department of Electrical & Computer Engineering
Georgia Institute of Technology
Atlanta, GA, USA

Preface

Wireless technology is revolutionizing the way we communicate and share information. The great advancements in wireless communications have been paralleled by advances in the computer industry, and the merging of the two promise a plethora of services for the consumer. A natural application for cellular radio networks is that of radiolocation in which the mobile unit is positioned based on measurement of signals at several base stations, or by the mobile unit itself making measurements to determine its own position. A large number of applications, which can benefit both consumers and service providers, stand ready to take advantage of location technology. The driving force behind the effort being expended to develop accurate location technologies today stems from the requirements imposed by the FCC in 1996. Wireless carriers have until October, 2001, to deploy a location system capable of locating a wireless 911 call to within 125 m. This remains the present target for accuracy, while in the future higher accuracy will most likely be required. High accuracy will also be required for some applications.

To address the issue of accurate location in CDMA cellular radio networks, we began research at the Georgia Institute of Technology which led to my doctoral dissertation. The original goal of the project was to identify and study the feasibility of using cellular networks to provide positioning information for Intelligent Vehicle/Highway Services[1] (IVHS). The FCC requirements provided further application for our research. The research sought to identify location methods for CDMA systems and examine their performance in realistic environments. We then began looking into the problems encountered in mobile radio systems, par-

[1]IVHS has been renamed to Intelligent Transportation Systems (ITS).

ticularly the interference that arises from the propagation channel and other user interference.

This book differs from my doctoral dissertation in a few ways. First, a few of the chapters have been rearranged and combined. Secondly, additional background material has been added in order to achieve a more comprehensive work. The background material in the introduction has been slightly lengthened and the propagation models of Chapter 3 have been expanded to include more of a mathematical description. Models for different scattering environments have also been included. A completely new chapter, Chapter 4, has been added to provide a thorough overview of methods for measuring location parameters, such as angles or times of arrival, and the algorithms that have been devised to use those parameters to form a location estimate. While there are many algorithms that can be found in the literature, a basic few receive the focus of the chapter while references for more information are included for further study by the reader. The multiple-access interference and non-line-of-sight propagation material of my dissertation have each been combined into two chapters with each providing analysis of the problems and methods for mitigating them. Another completely new chapter, Chapter 8, has been added in order to provide an overview of methods that are available within current cellular radio systems such as IS-95 CDMA and GSM. While the focus of the book is on CDMA cellular radio systems, much of the material, with the exceptions of Chapters 5 and 6, are applicable to other systems as well.

I hope the resulting book provides a useful introduction to the technologies and limitations of wireless location technologies in CDMA communications systems. Hopefully, both students and researchers will find the information useful, in the very least as a starting point for further investigation.

Acknowledgements

This book would not have been possible without the tremendous support and encouragement I have received. Most importantly, I wish to thank God for giving me this opportunity and providing me with the tools necessary to complete this endeavor. I also wish to share this work with my family for their encouragement and support, and with my wife who has been very understanding and supportive.

I wish to express my sincere gratitude and appreciation to my advisor at Georgia Tech, Professor Gordon L. Stüber, for providing me the opportunity to be part of his group and for his guidance, motivation, and support throughout the doctoral program. I am also grateful to my thesis committee members: Professors S. W. McLaughlin, M. A. Ingram, E. I. Verriest, and D. A. Klain for their time and valuable comments and suggestions relating to this research.

I am also pleased to acknowledge the support of GTE Laboratories, Incorporated, who funded the early part of this research, and the School of Electrical and Computer Engineering at the Georgia Institute of Technology for providing me this opportunity.

JAMES J. CAFFERY, JR.

To Jenny,
for your support,
friendship, and undying
love.

To my family,
without whose support
none of this would have
been possible.

Chapter 1

INTRODUCTION

Over the last decade, the deployment of wireless communications has been significant with an annual increase of cellular subscribers in the world averaging about 40% [103]. Currently, it is estimated that there are between between 36-46 million cellular users in the U.S. alone [51, 168], representing over 20% of the U.S. population. In the next few years, it is expected that a total of about 200 million wireless telephones will be in use worldwide [51], and that in 10 years, the demand for mobility will make wireless technology the primary source for voice communication, with a total market penetration of 50-60% [103]. In order to provide both an increase in capacity and better quality communications, the wireless industry has begun to migrate from the analog cellular networks that were first introduced in the early 1980's, to digital networks such as time-division multiple access (TDMA) and code-division multiple access (CDMA) in the U.S., Global System for Mobile communications (GSM) in Europe and Personal Digital Cellular (PDC) in Japan [32, 51, 119]. CDMA is a very popular choice due to its claims of higher system capacity, ability to mitigate multipath fading and other-user interference, universal frequency reuse, low transmission power, soft handoff capability, message privacy and the ability to exploit voice activity cycles [51, 94, 126, 127, 149, 175, 191]. CDMA assigns each user a unique pseudonoise (PN) spreading code, unlike frequency-division multiple access (FDMA) where each user is assigned a different frequency and TDMA where each user is assigned a different time slot.

As the move is made from analog to digital technology in second and third generation cellular and PCS systems, improved functionality and a broader spectrum of services will be made available to both system operators and consumers alike. The promise of even more enhanced services comes with the development of IMT-2000 (International Mobile Telephone 2000), the standardization process for which should be completed by the end of 1999 or early 2000. One of these new services, wireless position location, has received considerable attention over the past few years due to the Federal Communication Commission's (FCC) 1996 report and order [22]. The FCC mandate requires that all wireless service providers, including cellular, PCS, and SMR licensees, provide location information for Enhanced-911 (E-911) safety services, the provision for which requires two phases of completion. Phase I, the deadline for which passed in 1998, required that the wireless carriers relay a caller's telephone number and the location of the cell site and/or sector receiving the call to a designated Public Safety Answering Point (PSAP). Such capabilities allow the PSAP to call back if the call is disconnected. Phase II, to be completed by October 2001, requires that wireless carriers be able to report the location of all E-911 callers with an accuracy of 125 m (410 ft) in 67% of the cases. A further accuracy requirement of 13 m (40 ft) in 90% of the cases is possible for future mandates by the FCC [22].

1. APPLICATIONS

The provision of location information for wireless E-911 calls permits rapid response in situations where callers are disoriented, disabled, unable to speak, or do not know their location. In fact, a report by the FCC indicated that over 20% of 911 calls were made by wireless users and that one-fourth of those callers could not identify their location [22]. To compound the situation, the emergency operator receives very little network-based information regarding the location of the wireless caller. This is alarming considering that an increasingly large fraction of E-911 calls are placed by cell phones which is a direct result of the growing number of cellular and PCS subscribers. In 1994, approximately 50,000 wireless E-911 calls per day were made in the U.S., a figure that increased to 60,000 in 1996. By the year 2000, it is estimated that this figure will grow to 130,000. A 1996 study by the State of New Jersey indicated that wireless E-911 calls accounted for 43% of all calls to E-911 received during wireless location trials [168]. While the E-911 requirements of the FCC are the driving factor toward accurate location technology in the United States, outside of the states, there is also an interest for emergency services and for other services which represent the potential for large revenues.

While position location using radio signals has been used by the military for years, it is a relatively new application in civilian radio systems. The military has expended great effort to build a highly accurate location system for its various branches, but civilian industry has a more difficult time justifying the tremendous cost of developing such a system. Consequently, with the E-911 mandate of the FCC, research has focused on location methods which utilize the existing and future wireless system communications components, i.e., base stations (BSs) and mobile stations (MSs).

Besides its use for emergency management, location technology promises additional services. It will be the enabling technology for location sensitive billing, improved fraud detection, improved traffic management and many other services which have a large revenue potential for system operators. It is estimated that the total annual revenue potential for all location-based services will be over $8 billion dollars per year [178].

Location sensitive billing, fraud detection, system design, fleet management and Intelligent Transportation Systems (ITS) will benefit from wireless location technologies. Location sensitive billing will allow system operators to maximize profits and encourage usage behaviors by offering different rates depending on whether the phone is used at home, in the office, on the road, etc. [171]. This also provides wireless carriers with the opportunity to offer rates which will bring new subscribers into their customer base. Moreover, a service provider who may have multiple agreements with PCS, cellular, or satellite carriers, could offer its customers the ability to choose a carrier that best suits their needs at any given time and location [54].

Additionally, location technology can play a key role in the ongoing battle against cellular phone fraud. Annual industry fraud ranges on the order of $500 million, all of which is passed on to wireless customers in the form of higher phone usage rates. Some carriers estimate that up to 1% of their customer base experiences fraud each month. With the use of wireless location systems, it will be easier to find and catch the perpetrators.

A further application of location technology will be found in wireless system design and for radio resource management [62, 122]. With the ability to locate wireless calls and match them with their serving cell sites and received power levels, system planners could dramatically improve their ability to architect cells and wireless systems. More effective resource management could be obtained through the allocation of channels based on the knowledge of the wireless caller's location. Similarly, location information could assist in handoff procedures as the MS moves from one cell to another. Thus, cells could be better positioned and tuned, spectral efficiency improved and resource management made more effective through the use of accurate location technology.

In wireless networks, the system employs various means of tracking the relative location of mobile terminals since the location of a MS must be identified before a call to the MS can be established. The "location area" of a MS in the wireless network is often maintained in databases which are updated as the MS moves throughout the network. The use of exact location technology reduces and possibly eliminates the need for database management and reduces the system resources required since the MS can be paged in the cell in which it currently resides rather than paging several cells to identify its location.

Wireless location technology can also be used to make fleet operations more efficient and effective [84]. Having knowledge of the vehicles' locations allows a dispatcher to locate the nearest available vehicle, greatly improving response times. The improvement in field service not only applies to delivery operations, but to police and emergency vehicles as well as taxi and other service operators. Another developing technology that will rely heavily on accurate location information of mobile terminals is of Intelligent Transportation Systems. The strategic plan submitted by ITS America [79] calls for functionality in several areas, the two most important of which are listed below:

- Advanced Traffic Management Systems (ATMS) use various technologies to manage traffic in the transportation network.
- Advanced Traveler Information Services (ATIS) will provide information such as traffic safety alerts, routes of travel, and real-time traffic information directly to the traveler.

Route guidance, an ATIS service, will provide directions to travelers based on their known (estimated) current position. This service will inform the driver of the best route of travel based on traffic conditions, such as traffic congestion in a particular area. Furthermore, traveler information services will provide a "digital yellow pages" to give travelers the ability to locate such conveniences as nearby gas stations and hotels. References [33] and [204] provide a detailed overview of further uses of positioning technology for future transportation systems.

The author's previous research in wireless position location for CDMA networks showed that research into wireless position location leads to a diversity of research for supporting technologies. Position location not only requires efforts in the area of algorithm development but also in the technologies which provide estimates of location parameters such as times of arrival (TOAs), time differences of arrival (TDOAs) and angles of arrival (AOAs). Therefore, any research in wireless position location will necessitate the advancement of the supporting technologies required to produce accurate parameter estimates.

2. HISTORY OF WIRELESS LOCATION SYSTEMS

Interest in locating wireless radios can be traced back several decades. Over that time span, several location technologies have been developed and commercially deployed including Decca, Loran, Omega, the Global Positioning System (GPS) and the Global Navigation Satellite System (GLONASS). These technologies seek to provide location information for world-wide navigation.

The Decca system is designed to be a low-frequency (30-300 kHz) hyperbolic navigation system using continuous-wave comparison [131]. Phase comparisons are made between "master" and "slave" signals at a comparison frequency obtained from frequency multiplication of the master and slave signals. The master and slave stations are typically placed in a "Y" configuration, with the master at the center, and typical master-slave separations are on the order of 50-75 nautical miles (92-139 km). The phase difference measurements represent a hyperbolic line of position (see Chapter 3) which is used to identify the location of receiver. The Decca system is able to deliver location accuracies near 50 m with a range around 400 km [116].

The Loran (Long Range Navigation) positioning system was originally developed during World War II as an aid to the navigation of allied aircraft and the North Atlantic convoys [39]. Following the war, it was used by the US Coast Guard to aid marine navigation. This system was known as Loran A. Loran A evolved to the Loran C system to provide greater range and more precise, repeatable accuracy. Loran C is a hyperbolic system that uses low-frequencies (90-110 kHz) with an operating range in excess of 1000 nautical miles (1850 km). During its operation, Loran C is composed of 30 transmitting stations in several chains worldwide. From Loran C was developed a tactical version designed for short-range service known as Loran D. The advantage of Loran D over Loran C was the portability of the ground (transmitting) stations which allowed them to be quickly deployed. The mobility was gained at substantially reduced transmission power due to smaller transmitters and lower antenna towers. Location errors in Loran-C can run as high as 500 m.

The Omega navigation system is a long-range radio system developed by the United States Navy which provided full-time worldwide coverage from eight strategically located terrestrial very-low-frequency (10.2-13.6 kHz) transmitting stations [128]. It, too, is based on the hyperbolic location concept. This system has been utilized primarily by maritime and aeronautical users. The Omega system provides ranges up to several thousand kilometers, but with location accuracies only on the order of 1-3 km.

As these location systems illustrate, there is a trade-off between range and accuracy, both being functions of the carrier frequency. Microwave systems offered the highest accuracy (1 m), but can only operate within line-of-sight (LoS). Ultra-high frequency down to medium-frequency systems offered high accuracy (a few tens of meters) with ranges up to 400 km. The low-frequency systems offered greater range, but with much reduced accuracy (50 m). The very-low-frequency systems had the greatest range (several thousand kilometers) but with very poor accuracy (kilometers). These reasons illustrate the choice of transmission frequency used for the location systems discussed above.

Unlike the terrestrial-based location systems just mentioned, the global positioning system was conceived in the early 1970s as a highly accurate satellite-based navigation system. GPS, originally developed for military applications, is perhaps the most popular commercial location system today with applications for most forms of transportation and offering accurate location estimates when a LoS path exists to several satellites. GPS includes the Standard Positioning Service (SPS) which provides civilian users with 100 m accuracy and the Precise Positioning Service (PPS) which provides sub-20 m accuracy to military users [36]. Differential GPS provides 2-10 m accuracy to users within 1000 km of a fixed GPS reference receiver.

A location technology that uses signals from commercial FM radio stations has also been developed [28]. This technique uses pilot tones from FM stations, which are generally in in the 19 kHz range, to calculate positions using a reference station, similar to that done in differential GPS. Triangulation is performed to locate the MS using range estimates obtained from phase measurements. This technology has the advantage of wide coverage due to the high concentration of FM stations in many countries.

As the discussion above indicates, most of these technologies require that specialized equipment be placed in the MS, such as a GPS receiver, in order for location to be calculated in the MS. If the position needs to be known for ITS services or E-911, the location must be relayed by the MS to a central site, possibly over the cellular telephone network. An alternative is to use the wireless networks as the sole means to provide location services. The MS can be located by measuring the signals traveling between a MS and a fixed set of BSs. This approach takes advantage of the large pool of existing handsets with no modification and requires no other specialized equipment in the MS. Any necessary modifications, such as specialized location equipment, can be placed on the network side. We will address this issue further in Chapter 3.

As we will see in the remaining chapters, the method of radiolocation, one of the many methods available for computing location estimates, proves to be the most suitable for wireless networks. Unfortunately, the harsh propagation conditions in wireless networks present a formidable challenge for implementing and deploying accurate location systems. Multipath fading produces errors in the parameters that are measured and used for location. When there is no direct path between the MS and a BS, a MS's signal arrives at a BS from a path that is longer and from a different direction than the direct or LoS path due to reflection and/or diffraction which results in location measurement errors. Finally, as with all cellular systems, other-user interference will introduce error into the location measurements. In CDMA, power control schemes are used to combat the near-far effect; however, it is still a factor for reverse link location (MS to BS) systems since the signals from a MS must be measured at multiple BSs. For forward link location (BS to MS), the same problem applies since the MS must measure multiple interfering pilot signals of unequal power from nearby BSs.

3. OVERVIEW

This book explores the the performance of wireless position location in cellular radio systems and seeks to identify solutions for the major impairments found in such systems. A performance evaluation of conventional wireless location approaches is necessary as a starting point to determine the feasibility of meeting accuracy requirements such as those proposed by the FCC. The propagation environment introduces, among other things, multiple-access interference and LoS obstructions. Multiple-access interference (MAI) results from users sharing the same bandwidth in CDMA systems. As CDMA systems have been shown to be interference limited [104], MAI will limit not only the accuracy to which parameters (delays, angles, etc.) that are necessary for location can be made, but also whether they can be made at all because of the near-far effect. Non-line-of-sight (NLoS) propagation, perhaps, poses the greatest deterrent to accurate location, since a direct path between the MS and BS rarely exists in a cellular environment. In the following, a brief description of the topics and challenges are presented.

3.1 PERFORMANCE IN CDMA SYSTEMS

To avoid adding additional equipment to the MS handset and allow existing handsets to be located, CDMA cellular radio networks have been proposed as the sole means of providing wireless location services. Hence, it is necessary to determine the accuracy that can be achieved in a typical cellular system, and the effects that varying propagation conditions have on performance. Previous location studies in cellular networks have concentrated on signal strength [63, 139, 161], angle of arrival (AoA) [144] and time of arrival (ToA) [55, 62, 65] methods. However, the ToA radiolocation investigations in [55] and [62] assumed only multipath propagation effects in macrocells where LoS propagation was assumed between the MS and a BS. The work in [65] used the ToA method with the location estimate computed at the MS and also assumed LoS propagation.

We seek to determine the performance of time-based and direction finding radiolocation techniques in CDMA systems using conventional receiver structures. The methods are evaluated for ranging (one-dimensional) and position (two-dimensional) location in both macrocells and microcells. A conventional time delay estimation and tracking method using the delay lock loop (DLL) is employed to provide the ToA estimates to the location algorithm. The effect of the number of BSs used for location and the varying propagation channel conditions are investigated. Furthermore, techniques are developed to improve the location accuracy.

3.2 EFFECTS OF MULTIPLE-ACCESS INTERFERENCE

In CDMA cellular radio systems, users share the same frequency band and are separated using different spreading codes. Due to the lack of code orthogonality, interference arises amongst the users. This has important implications for location systems that use conventional tracking techniques, such as the DLL, to provide time delay estimates for location algorithms. Since several BSs are required to form a location estimate, the signal from the MS must be detected and the time delays tracked at several BSs. While power control is designed to mitigate the near-far effect within a MS's own cell, it is still a problem at other BSs since the MS is not power-controlled to those BSs. The inability to detect or track the time delays with high accuracy at several BSs directly affects the accuracy to which a location estimate can be made. Consequently, the performance of the non-coherent DLL (NC-DLL) is rigorously analyzed to determine the affects of MAI.

Only one previous analysis of the DLL in MAI has been given in the literature [70]. This work is founded on the assumption that the MAI can be accurately approximated as Gaussian noise. However, the Gaussian assumption is not always accurate for short spreading codes, when there are few interferers, or when the signal-to-noise ratio (SNR) is high [200]. Also, it is not valid when a single user dominates the multiple access interference [98]. In contrast to [70], the analysis presented in this book uses a simple, but accurate, approximation for the aperiodic cross-correlation function (ACF) between two spreading sequences in order to simplify the computation of higher order moments that are involved in the analysis.

As the analysis indicates, the interference from other users has a drastic effect on the accuracy to which time delay estimates can be made. Not only is the tracking performance reduced, but it becomes difficult for nearby BSs, other than the serving BS, to detect the desired user's signal even with perfect power control. To overcome the near-far effect, multiuser detectors and parameter estimators have been proposed. Near-far resistant estimators for constant or slowly varying channel parameters have been developed based on signal subspace techniques [8, 174]. These have been extended to the case of fading channels, but again with constant or slowly varying time delays [173]. In addition, joint multiuser detectors and parameter estimators have been developed [78, 133]. For estimating time-varying delays, an extension of the subspace techniques has been proposed [81], while time delay trackers based on the DLL with MAI cancellation have also been developed [89, 205].

As an extension to these developments, this book develops a multiuser parameter estimator for joint estimation and tracking of *time-varying* amplitudes and delays. The estimation problem is approached from a system identification point of view where a nonlinear adaptive filter is used to approximate the output of the channel. A nonlinear extension of the standard linear Kalman filter (KF) [83] is used that avoids linearization and has less complexity than the extended Kalman filter (EKF).

3.3 EFFECT OF NLOS PROPAGATION

One of the major sources of error for time-based radiolocation in cellular networks is NLoS propagation [16, 154, 197]. The algorithms that have been developed for location are always based on the assumption that a direct, or LoS, path exists between the MS and each BS that is used for location. Unfortunately, a LoS path rarely exists between a MS and its serving BS, let alone nearby surrounding BSs in macrocells. In microcells, because of the smaller size of the cells, it is usually assumed that a LoS path exists to the serving BS, but not to the other BSs. Consequently, the surrounding BSs will receive the desired signal from reflections and/or diffractions, not from a direct path. The measured time delays, then, will reflect a distance that is greater than the true distance between the MS and BS. The extra propagation distance is introduced into the location algorithms as a positive bias that is added to the true value. To quantify the effect of NLoS propagation, several location algorithms are analyzed and compared when their time delay measurements are corrupted by the measurement bias.

The measurement bias introduced by NLoS propagation causes the location algorithms to form a location estimate that can be far from the true location, depending on the size of the bias. Since the NLoS bias is present for at least one BS involved in the location process, it must removed or accounted for in order to obtain accurate location estimates in cellular systems. One method has been proposed that can adjust a series of range estimates, taken over a period of time, to approximately their true LoS values [197]. This method implicitly subtracts the NLoS bias from its measurements, but unfortunately, it assumes that the statistics of the NLoS bias and the measurement noise are known.

This book describes and evaluates those techniques which attempt to reduce the effect of the bias and produce location estimates nearer the true location. The method just described operates by reconstructing the LoS measurements from the NLoS corrupted measurements. Another method to be presented weights the contribution of the NLoS corrupted measurements to reduce their influence on the location estimate.

4. OUTLINE

This book is organized as follows. Chapter 2 develops the signal and channel models that are used throughout the remaining chapters. The channel models incorporate fading, shadowing and path loss. Scattering models for different environments are also presented. The signal model is based on CDMA signaling in multipath fading and multiple-user interference. A brief review of different location methods (dead-reckoning, proximity detection and radiolocation) is presented in Chapter 3. The major channel impairments that limit radiolocation accuracy are also discussed. Chapter 4 presents various methods of obtaining measurements of the location parameters, such as angles or times of arrival, for the different forms of radiolocation. Several algorithms that use these location parameters to estimate the location of an MS are derived and discussed. In Chapter 5, the performance of a time-based wireless location system in CDMA cellular networks is evaluated. The effects of the propagation environment, number of BS used for location, and NLoS propagation are also investigated. Chapter 6 presents an analytical study of the effect of MAI on the performance of the non-coherent delay lock loop. Methods to mitigate the effects of MAI are discussed and evaluated. The effect of NLoS propagation on location accuracy is studied in Chapter 7 with the analyses of several location algorithms with biased measurements. Two techniques for mitigating its effect are presented. Chapter 8 provides an overview of the capability for wireless location that is provided in current and future cellular and PCS communications systems. Finally, Chapter 9 summarizes the results from each of the previous chapters and addresses several topics for future research.

Chapter 2

SYSTEM AND CHANNEL MODELS

This chapter provides background information that will be used in later chapters of the book. Models for the mobile radio propagation channel and for CDMA communications signaling are presented with emphasis on the underlying mathematical models.

1. CHANNEL PROPAGATION MODEL

The mobile radio channel places fundamental limitations on the performance of wireless communications systems and will play an important role in location systems that are based on cellular radio communications. The transmission path between the MS and BS can vary from simple LoS to one that is severely restricted by buildings, hills, or foliage. The channel characteristics are also dependent on the deployment used (macrocells, microcells, or indoor picocells). In macrocells, the BS antennas are elevated well above the terrain and away from scatterers while the MS is surrounded by local scatterers [47]. Hence, a direct LoS path between the MS and it serving BS is unlikely. In microcells, a LoS path is often assumed to exist to its serving BS [1, 57]. In both deployment schemes, it is unlikely that a LoS path will exist to BSs other than the MS's serving BS. Generally, the mobile radio propagation channel is characterized by three nearly independent phenomena: multipath fading, shadowing and path loss [169, 121, 175]. A brief description of each of these phenomena follows along with a presentation of their mathematical models. Scattering models are also discussed.

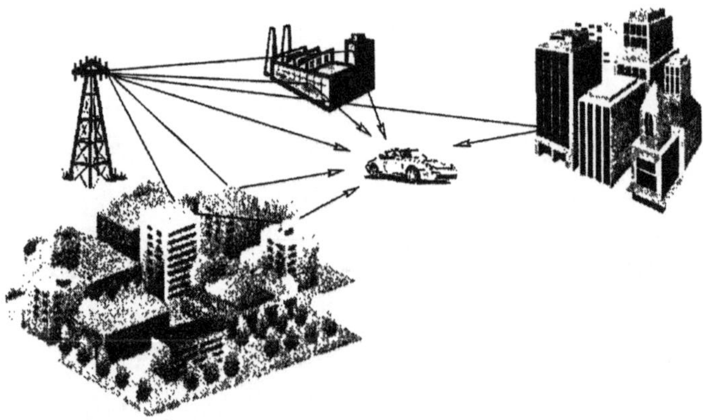

Figure 2.1. Multiple reception of a signal from reflected and diffracted paths leads to the multipath fading phenomenon.

1.1 MULTIPATH FADING

Multipath fading, also called fast fading, is the rapid fluctuation of the complex envelope of the received signal caused by reception of multiple copies of a transmitted signal, each with different amplitude, phase and delay. Fig. 2.1 illustrates the concept of multipath propagation. The reflected signals arriving at the MS or BS will add constructively or destructively resulting in fades as deep as 30 dB when the MS moves only a fraction of a wavelength [121]. Spread spectrum systems provide some immunity to multipath fading since paths that are separated by a chip period are essentially uncorrelated [67, 94, 127].

In macrocells, non-isotropic scattering is assumed since the MS is surrounded by local scatterers and the signals arrive from many different angles with no dominant component [47, 80, 175]. As a result, the amplitude distribution of multipath fading is described by a Rayleigh distribution [13, 121, 156]. In microcells, the amplitude distribution is often described by a Rician distribution due to the presence of a strong LoS or specular path between the MS and BS [14, 57]. A consequence of the LoS path is decreased fading depth [169]. An important parameter for Rician distributions is the *Rice factor* which is the ratio of the power in the LoS or specular component to the power in the scattered components. When the Rice factor is zero, the distribution is Rayleigh, and as the factor becomes large, the distribution approaches a Gaussian. Values for the Rice factor have been found experimentally to range from 5 to 30 dB [14, 56, 169].

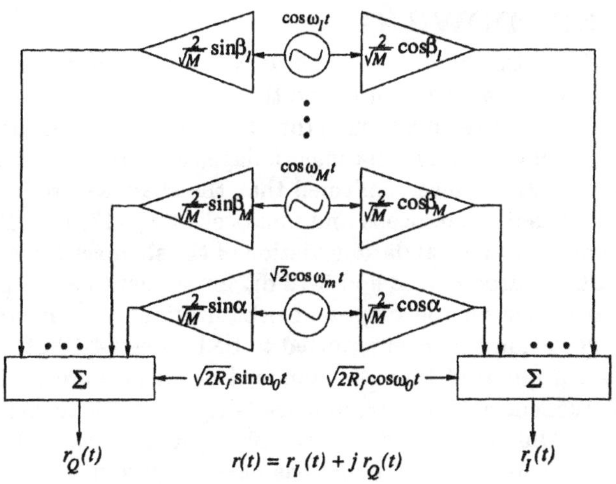

Figure 2.2. **Jakes' multipath fading simulator incorporating the Rice factor, R_f, for Rician fading.**

The development of the system model in Section 2 of this chapter will assume a wideband channel model with M taps. For numerical convenience, the tap delays can be chosen as integer multiples of some delay τ, i.e., $\tau_i = i\tau$, for some integer i. The multipath coefficients of the channel, $f_{k,i}(t)$, can be generated using Jakes' fading simulator [80], which provides taps with the appropriate distributions (i.e, Rayleigh or Rician) and near the correct tap autocorrelations, although the taps are somewhat correlated. Jakes' simulator, originally developed for Rayleigh fading channels, is extended here for Rician fading channels as shown in Fig. 2.2. The original simulator is modified to make use of Aulin's Rician fading model [3] where the phase and quadrature components of the LoS signal are given by

$$m_I(t) = s \cdot \cos\omega_0 t \tag{2.1}$$
$$m_Q(t) = s \cdot \sin\omega_0 t \tag{2.2}$$

where $s^2 = 2R_f$. This value of s arises since the simulator has been modified for unit power in the scatter component.

The statistical properties of Jakes' simulator have been derived in [123]. Jakes' model has also been extended to incorporate the angular spread due to a disk of scatterers on an antenna array [45].

1.2 SHADOWING

Shadowing, also referred to as slow fading or shadow fading, manifests itself as a slow variation in the the mean envelope over a distance corresponding to tens of wavelengths. It is caused by variations in the local topography such as buildings, foliage, and hilly terrain. Experimental observations have confirmed that the shadows are log-normally distributed in both macrocells and microcells [93, 117, 121, 138].

For macrocells, the standard deviation of the shadows typically ranges from 5-8 dB for urban areas and 8-12 dB for suburban areas [138] showing a dependence on the degree of urbanization. For microcells, the standard deviation has been reported to be between 4-13 dB [9, 107]. It has also been reported that the standard deviations are slightly larger at higher frequencies [110, 113] and are fairly invariant to changes in BS height [113]. Only a few studies have investigated the spatial correlation of shadows [59, 107], and a model has been suggested where shadowing is modeled as a Gaussian white noise process that is filtered by a first order low-pass filter [59]. This simple Markovian model attempts to describe variations in the local mean envelope (or squared envelope) level due to shadow variations and to account for the spatial correlation of the shadows. With this model,

$$\Omega_{k+1(\mathrm{dB})} = \varsigma\Omega_{k(\mathrm{dB})} + (1-\varsigma)\eta_k \tag{2.3}$$

where $\Omega_{k\ (\mathrm{dB})}$ is the local mean envelope (or square-envelope) level (in dB) that is experienced at location k, ς is a parameter that controls the spatial decorrelation of the shadowing, and $\{\eta_k\}$ is a zero-mean discrete-time Gaussian random process with autocorrelation $\phi_{\eta\eta}(l) = \sigma_\eta^2\delta(l)$. The autocorrelation of $\Omega_{k\ (\mathrm{dB})}$ is given by

$$\phi_{\Omega_{(\mathrm{dB})}\Omega_{(\mathrm{dB})}}(l) = \frac{1-\varsigma}{1+\varsigma}\sigma_\eta^2\varsigma^{|l|} = \sigma_s^2\varsigma^{|l|} \ . \tag{2.4}$$

where σ_s is called the shadow standard deviation. If we assume that the local mean is sampled every T_s seconds, then the autocorrelation can be expressed as

$$\phi_{\Omega_{(\mathrm{dB})}\Omega_{(\mathrm{dB})}}(k) \equiv \phi_{\Omega_{(\mathrm{dB})}\Omega_{(\mathrm{dB})}}(l = kT_s) = \sigma_s^2\epsilon_D^{(vT_s/D)|k|} \tag{2.5}$$

where ϵ_D determines the correlation between two points separated by a spatial distance D and v is the velocity of the MS. This model can be used for both macrocells and microcells.

1.3 PATH LOSS

Path loss predicts how the signal power decays with distance from a BS. An inverse fourth-law power loss with distance is often used to characterize path loss in macrocells [80, 121]. Hata [63] developed a useful model for path loss in macrocells based on the experimental results of Okumura [117]. The model expresses the path loss as a function of BS height, MS height, carrier frequency and the type of environment (urban, suburban or rural). With Hata's model, the path loss is expressed as (in dB)

$$L_p(d) = \begin{cases} A + B\log_{10}(d) & \text{urban area} \\ A + B\log_{10}(d) - C & \text{suburban area} \\ A + B\log_{10}(d) - D & \text{open area} \end{cases} \tag{2.6}$$

where

$$A = 69.55 + 26.16\log_{10}(f_c) - 13.82\log_{10}(h_b) - a(h_m) \tag{2.7}$$

$$B = 44.9 - 6.55\log_{10}(h_b) \tag{2.8}$$

$$C = 5.4 + 2\left[\log_{10}(f_c/28)\right]^2 \tag{2.9}$$

$$D = 40.94 + 4.78\left[\log_{10}(f_c)\right]^2 - 19.33\log_{10}(f_c) \tag{2.10}$$

and

$$a(h_m) = \begin{cases} (1.1\log_{10}(f_c) - 0.7)h_m - (1.56\log_{10}(f_c) - 0.8) \\ 8.28(\log_{10}(1.54h_m))^2 - 1.1 \\ 3.2(\log_{10}(11.75h_m))^2 - 4.97 \end{cases} \tag{2.11}$$

for a medium/small city, a large city with $f_c \geq 400$ MHz, and a large city with $f_c < 400$ MHz, respectively. Also, $450 \leq f_c \leq 1000$ MHz is the carrier frequency, $30 \leq h_b \leq 200$ m is the BS height, $1 \leq h_m \leq 10$ m is the MS height and $1 \leq d \leq 20$ km is the distance between the MS and BS.

Another well known model was developed by Lee [96]. However, Lee's model is generally used to predict path loss over flat terrain, so caution must be taken when using it to model hilly terrain.

In microcells, the path loss behavior is different. For outdoor microcells, two models have resulted from the European COST231 study [31]: the COST231-Hata and COST231-Walfish-Ikegami models. For urban street microcells, several authors have found that the path loss follows a two-slope characteristic [61, 12, 56, 38]. Using this model, the signal power attenuates by an inverse square law with distance from the BS up to distance called the *breakpoint*. After the breakpoint, the signal power attenuates with an inverse fourth power law. Mathematically, the area mean $\bar{\Omega}(d) = \mathrm{E}[\Omega(d)]$ is given by [61]

$$\bar{\Omega}(d) = 10\log_{10}\left(\frac{A}{d^{m_a}(1 + d/g)^{m_b}}\right) \quad \text{dBm} \tag{2.12}$$

where A is a constant, d is the radio path length, g is the breakpoint, and m_a and m_b determine the slopes before and after the break point. The breakpoint has been reported to range between 150 and 300 m [61, 198].

Street microcells may also exhibit NLoS propagation when a MS rounds a street corner, a phenomenon known as the the *corner effect*. When this happens, the signal power can drop 20-30 dB over relatively short distances [143, 182]. To account for this effect, LoS propagation is assumed to the MS until it rounds the corner. The NLoS propagation after rounding a street corner is modeled by assuming LoS propagation from an imaginary transmitter that is located at the street corner having a transmit power equal to the received power at the street corner from the serving BS. The area mean (in dBm) is given by

$$\bar{\Omega}(d) \sim \begin{cases} 10\log_{10}\left(\frac{1}{d^{m_a}(1+d/g)^{m_b}}\right) & d \leq d_c \\ 10\log_{10}\left(\frac{1}{d_c^{m_a}(1+d_c/g)^{m_b}} \cdot \frac{1}{(d-d_c)^{m_a}(1+(d-d_c)/g)^{m_b}}\right) & d > d_c \end{cases} \tag{2.13}$$

where d_c is the distance between the serving BS and the corner.

1.4 SPATIAL SCATTERING MODELS

A scattering model will be useful in Chapter 5 for simulation of angles of arrival and times of arrival in NLoS propagation environments. The amount of scattering that is experienced at the MS and BS depends on the deployment used. In macrocells, the BS is placed high above local scatterers and receives signals over a narrow azimuthal spread. The MS is surrounded by local scatterers and thus receives signals over a very broad angular spread. One of the first models to describe this phenomenon consisted of a MS surrounded by a uniform ring of scatterers [47, 80], as illustrated in Fig. 2.3(a). A similar model in which discrete local scatterers were evenly spaced on a ring about the mobile was discussed in [91]. Recently, other models have been proposed to describe the location of the scatterers in the ring about the MS, while keeping the same basic approach. In [100], the ring of scatterers was replaced with a Gaussian distribution of the scatterers about the MS and a probability distribution function (*pdf*) for the AOAs was derived. A uniform disk of scatterers about the MS was considered in [45].

In microcells, the BSs may be placed below roof top level at lamp post height in order to confine the signal to small area [61, 175]. Consequently, the BSs will be surrounded by local scatterers just like the MS as shown in Fig. 2.3(b). Typically, ray tracing techniques are used in microcells to simulate the propagation channel. However, one model for this scattering environment has been developed that uses an elliptical

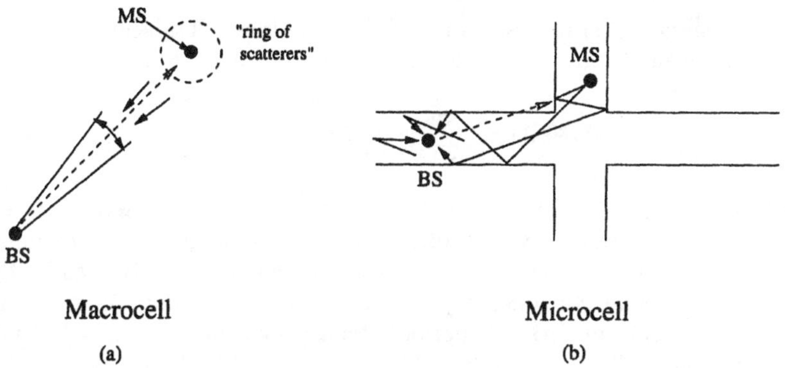

Figure 2.3. A depiction of scattering for (a) macrocells and (b) microcells in mobile radio propagation environments. The dashed lines indicate the true direction from the BS to the MS.

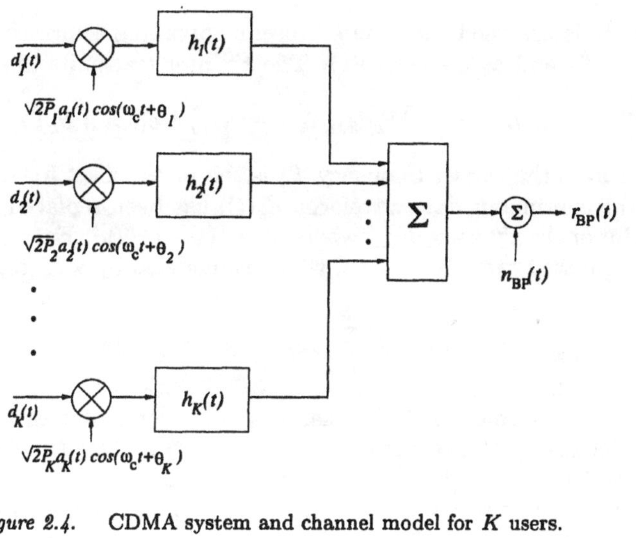

Figure 2.4. CDMA system and channel model for K users.

scattering model which encompasses local scatterers around the MS and BS [97].

2. SIGNALING SYSTEM MODEL

Throughout the rest of this book, a multiple-access CDMA communications system with K users is considered as shown in Fig. 2.4. In the following, a brief development of the communications signaling model for the multiuser CDMA channel is given.

The signal $d_k(t)$ represents the kth user's binary data sequence which is comprised of unit amplitude pulses of duration T_b seconds,

$$d_k(t) = \sum_{m=-\infty}^{\infty} d_{k,m} p_{T_b}(t - mT_b) \tag{2.14}$$

where $d_{k,m} \in \{-1, +1\}$ is the mth data bit of the kth user and $p_Y(t) = 1$ for $0 \leq t \leq Y$ and $p_Y(t) = 0$ otherwise. The data signals are assumed to be independent between users. Each data signal is modulated by the user's PN spreading code, $a_k(t)$, followed by carrier modulation. The spreading code consists of a periodic binary sequence of unit amplitude pulses of duration T_c seconds:

$$a_k(t) = \sum_{n=0}^{N-1} a_{k,n} p_{T_c}(t - nT_c) \tag{2.15}$$

where it is assumed that the kth user's spreading sequence has period $N = T_b/T_c$ and $a_{k,n} \in \{-1, +1\}$. The kth user transmits the signal

$$s_k(t) = \sqrt{2P_k}\, d_{k,m} a_k(t - mT_b) \cos(\omega_c t + \theta_k) \tag{2.16}$$

where ω_c is the carrier frequency, θ_k is the phase offset for the kth user, and the continuous date waveform, $d_k(t)$, has been replaced by the mth data bit of the kth user, $d_{k,m}$, where $m = \lfloor (t - \tau_k)/T_b \rfloor$. Each user's signal is transmitted over the channel which is modeled by a M-tap delay line

$$h_k(t) = \sum_{i=1}^{M} f_{k,i}(t) \delta(t - \tau_{k,i}(t)) \tag{2.17}$$

where it is assumed that the channels are different for each user. The coefficients, $f_{k,i}(t)$, are the tap weights and $\tau_{k,i}(t)$ are the delays of the multipaths associated with the kth user's channel.

The bandpass received signal is the sum of the users' delayed signals and is given by

$$\begin{aligned} r_{\mathrm{BP}}(t) &= \sum_{k=1}^{K} \sum_{i=1}^{M} \sqrt{2P_k}\, d_{k,m} f_{k,i}(t) a_k(t - mT_b - \tau_{k,i}(t)) \cos(\omega_c t + \phi_{k,i}(t)) \\ &\quad + n_{\mathrm{BP}}(t) \end{aligned} \tag{2.18}$$

where the $\phi_{k,i}(t) = \theta_k - \omega_c \tau_{k,i}(t)$ is the received phase of the kth carrier, and $n_{\mathrm{BP}}(t)$ is additive white Gaussian noise given by

$$n_{\mathrm{BP}}(t) = \sqrt{2} n_c(t) \cos(\omega_c t) - \sqrt{2} n_s(t) \sin(\omega_c t) \tag{2.19}$$

where $n_c(t)$ and $n_s(t)$ are zero-mean, independent Gaussian random process of variance $N_0/2$. Both $n_c(t)$ and $n_s(t)$ have the same power spectral density as $n_{\text{BP}}(t)$

Before the received signal is sampled and passed to the digital receiver, it is band-limited to a minimum of $1/T_c$ Hz. After down-conversion, this is done by an ideal rectangular low-pass filter having impulse response

$$h_{\text{LP}}(t) = \frac{\sin \frac{2\pi t}{T_c}}{\pi t} \ . \tag{2.20}$$

The result of the convolution of the down-converted $r_{\text{BP}}(t)$ and $h_{\text{LP}}(t)$ is

$$r(t) = \sum_{k=1}^{K} \sum_{i=1}^{M} \sqrt{P_k} d_{k,m} f_{k,i}(t) e^{j\phi_{k,i}(t)} \bar{a}_k(t - mT_b - \tau_{k,i}(t)) + n(t) \tag{2.21}$$

where

$$\begin{aligned} \bar{a}_k(t) &= \sum_{n=0}^{N-1} a_{k,n} \frac{1}{\pi} \left[\mathcal{S}i\left(2\pi \frac{t - nT_c}{T_c}\right) \right. \\ & \left. \qquad - \mathcal{S}i\left(2\pi \frac{t-(n+1)T_c}{T_c}\right) \right] \end{aligned} \tag{2.22}$$

is the low-pass filtered spreading code of the kth user and the function $\mathcal{S}i(x)$ is defined to be

$$\mathcal{S}i(x) = \int_0^x \frac{\sin(y)}{y} \, dy \ . \tag{2.23}$$

We assume that the received signal is sampled at q times the chip rate, giving a period between samples of $T_s = T_c/q$. Therefore, the received baseband signal sampled at $t = lT_s$ becomes:

$$r(l) = \sum_{k=1}^{K} \sum_{i=1}^{M} \beta_{k,i}(l) d_{k,m_l} \bar{a}_k(lT_s - m_l T_b - \tau_{k,i}(l)) + n(l) \tag{2.24}$$

where $\beta_{k,i}(l) = \sqrt{P_k} f_{k,i}(l) e^{j\phi_{k,i}(l)}$ incorporates the channel amplitude, $f_{k,i}(l)$, the transmitted power, P_k, and the received phase, $\phi_{k,i}$, of the kth user, and $m_l = \lfloor (lT_s - \tau_k(l))/T_b \rfloor$. The noise term, $n(l)$, is assumed to have zero mean and a variance of σ_n^2.

It will be convenient in later chapters to write the scalar sampled received signal, $r(l)$, in matrix form for a single tap channel ($M = 1$) as follows:

$$r(l) = \boldsymbol{\beta}^T(l) \boldsymbol{\mathcal{D}}(l) \bar{\mathbf{a}}(l) + n(l) \tag{2.25}$$

where

$$\boldsymbol{\beta}(l) = \begin{bmatrix} \beta_1(l) \\ \vdots \\ \beta_K(l) \end{bmatrix}, \qquad \boldsymbol{\mathcal{D}}(l) = \begin{bmatrix} d_{1,m_l} & & \mathbf{0} \\ & \ddots & \\ \mathbf{0} & & d_{K,m_l} \end{bmatrix},$$

$$\text{and} \quad \bar{\boldsymbol{a}}(l) = \begin{bmatrix} \bar{a}_1(lT_s - m_lT_b - \tau_1(l)) \\ \vdots \\ \bar{a}_K(lT_s - m_lT_b - \tau_K(l)) \end{bmatrix} .$$

Chapter 3

LOCATION METHODS AND IMPAIRMENTS

In this chapter, several methods for determining the location of a MS are presented. The methods range from accumulative travel estimators to proximity detectors to range and angle estimators. Each of these has advantages and disadvantages in their possible application of providing location services in cellular and PCS networks. In the final part of the chapter, impairments to location accuracy are discussed in detail along with possible solutions. Approaches for mitigating these impairments are the subjects of the later chapters.

1. LOCATION METHODS

Position location has been an active field of study over the last few decades. Originally, location methods were sought for the purpose of monitoring vehicles from law enforcement agencies, taxi fleet operations, emergency medical services, cargo fleets and public mass transit [122, 141]. The driving force behind the research was the desire to improve the field service of these operations. The most recent applications of location technology have been in mobile cellular radio for purposes of cellular system design, channel assignment, handoffs, determining cell service area, fraud deterrence, E-911 and traffic monitoring and management [22, 55, 62, 63, 118, 159, 161, 193, 196].

Simply put, location systems entail the acquisition of information about the location of a MS (or fleet of MSs) operating in an area and the processing of that information to form location estimates. They are usually distance or direction estimators, or both. Overall, the location systems that have been proposed can generally be placed in three categories: dead-reckoning, proximity systems and radiolocation. In the remainder of this section, we will provide an overview of these methods.

1.1 DEAD-RECKONING

Dead-reckoning is a primitive location technique that is based on the concept of computing the direction and distance of travel from a known starting position [141]. The system relies on accurate measurement of the MS's acceleration, velocity, and direction of travel. Several components can be used for the implementation of a dead-reckoning positioning system. These can include transmission, wheel and inertial sensors as well as magnetic compasses. Transmission sensors measure the angular position of the transmission shaft in order to formulate an estimate of the distance traveled and/or velocity of the MS. Similarly, wheel sensors can make use of existing sensors from an anti-lock breaking system (ABS) or differential odometry to provide estimates of distance traveled and heading change information [204]. Inertial sensors, such as gyroscopes and accelerometers, provide methods of measuring the position and velocity of the MS. Finally, magnetic compasses provide an inexpensive means of determining vehicle heading.

The sensors discussed above can be used to measure the direction the MS is traveling, θ, and the distance traveled, d. Given a known starting position[1], $\boldsymbol{x}_0 = [x_0, y_0]^T$, the update for the location at time n is given by the simple relation

$$\boldsymbol{x}_n = \boldsymbol{x}_0 + \sum_{i=0}^{n-1} \boldsymbol{d}_i \tag{3.1}$$

where $\boldsymbol{d}_i = [d_i \cos\theta_i, d_i \sin\theta_i]^T$ is the displacement vector. The quantities d_i and θ_i are the magnitude and the direction of the displacement vector between times $n-1$ and n. The location update process for dead-reckoning is illustrated in Fig. 3.1.

Dead-reckoning technology has been used with maps on CD-ROM to give direction to travelers [84]. Many factors including road conditions, weather conditions (such as strong side winds or slick roads), imprecise measurement equipment and even changes in tire pressure lead to position errors. Unfortunately in a dead-reckoning location system, the updated position depends on the previous estimates. Hence, errors tend to accumulate with distance. This problem may be alleviated by periodically updating the position of the vehicle by some other method and again using dead-reckoning until the next update [141]. Dead-reckoning has also been used in conjunction with map-matching methods to improve the accuracy of the location system [33].

[1] In the remainder of the book, lower-case bold variables will indicate vectors while upper-case bold letters will denote matrices. All non-bold characters will denote scalar values. The notation $[\cdot]^T$ denotes the transpose of a vector or matrix.

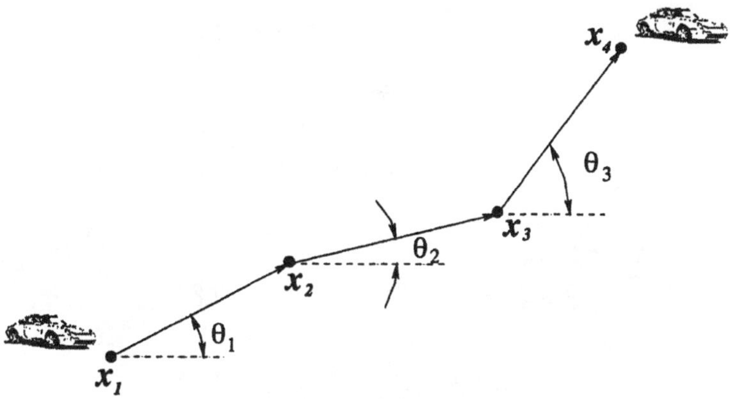

Figure 3.1. Location updates using the method of dead-reckoning.

1.2 PROXIMITY SYSTEMS

Proximity location systems, also known as signpost or beacon systems, are perhaps the conceptually simplest location systems. Location estimates are obtained through the principle of fixed reference steering. The vehicle's location is determined from its proximity to fixed detection devices, which can be anything from magnetic sensors to conventional radio transmitters and receivers [135, 141] (see Fig. 3.2). This method is very attractive for *fixed route* operations such as those performed by delivery vehicles or even for small cities. However, for larger areas, a prohibitively large infrastructure of devices is required. Proximity detection offers the capability of either self- or remote-positioning. With self-positioning, the beacon from the signposts must contain information regarding the position of the beacon. Likewise, identification of vehicles by the network for remote positioning requires that information be embedded into the beacon signals emanating from the signposts.

It is interesting to note that a cellular radio system could be used as a proximity location system. A very coarse estimate of the location of the MS could be obtained from its nearness to a BS or by the sector of the cell the MS is in (Fig. 3.3). Thus, the BS antennas would replace the specialized detectors shown in Fig. 3.2 providing a wide area of coverage for proximity detection. Of course, the usefulness of this approach as a location service and its general accuracy depends on the size of the cells.

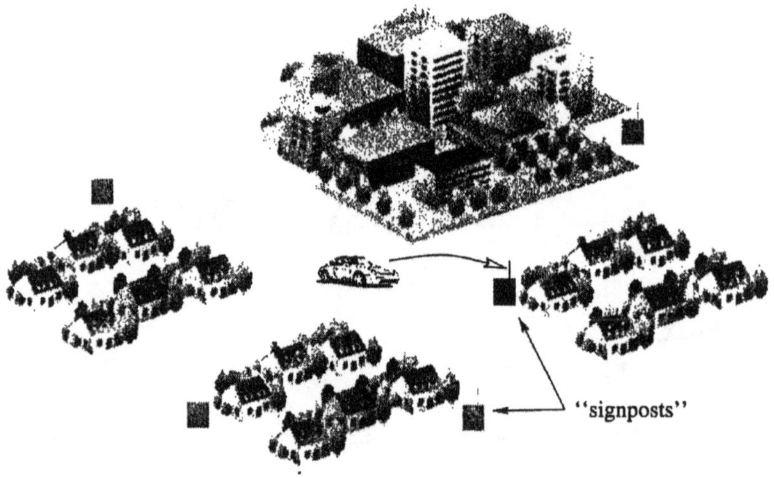

Figure 3.2. The nearness of the MS to the proximity detectors or signposts gives an indication of the location of the MS. The detector which has detected the vehicle nearby is solid black.

1.3 RADIOLOCATION

Radiolocation systems attempt to locate a MS by measuring the radio signals traveling between the MS and a set of fixed stations (FSs). With the commercial location technologies discussed previously (i.e., Loran, GPS), radiolocation is used in which the MS formulates its own position from signals received from the FSs. This form of location is often referred to as *self-positioning.* In these systems, a special receiver is placed in the MS which calculates the MS's position. Alternatively, the position of the MS could be calculated at a remote location based on signals received at the FSs. This form of radiolocation is known as *remote-positioning* and requires a transmitter for the MS.

In a cellular or PCS environment, the BSs serve the role of the FSs and the MSs are equipped with transmitting/receiving equipment. Radiolocation can be implemented on the forward link or reverse link. With forward link location, the MS uses signals transmitted by several BSs to calculate its own position (self-positioning). This is the approach used in GPS where satellites are used instead of terrestrial BSs. With reverse link location, several BSs measure the signals transmitted by the MS and relay them to a central site for processing (remote-positioning). For location in cellular networks, the second approach has the advantage of not requiring any modifications or specialized equipment in the MS

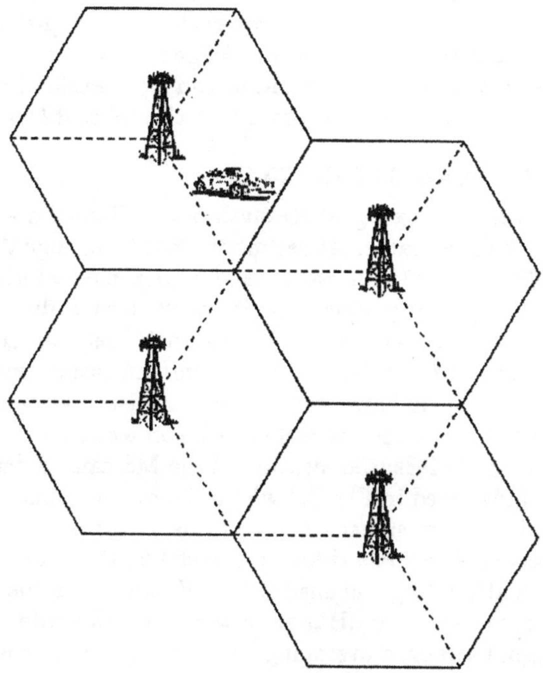

Figure 3.3. A cellular network operating as a proximity detection (or signpost navigation) system. The gray sector which is communicating with the MS provides a coarse estimate of its location.

handset, thus accommodating the large pool of handsets already in use in existing cellular networks.

There are three fundamental types of radiolocation systems: those based on signal strength, angle of arrival (AoA), or time (ToA or TDoA) measurements [41, 42, 47, 62, 63, 118, 144, 161]. The signal measurements are first used to determine the length or direction of the radio paths to/from a MS from/to multiple BSs, and then known geometric relationships are used to determine the location [141]. In order to achieve high accuracy in location estimates, it is necessary that line-of-sight (LoS) paths exist between the MS and the BSs that are utilized in the location process.

In general, locating a MS in two-dimensions requires a minimum of three BSs in order to resolve ambiguities arising from multiple crossings of the lines of position. As we will see in the following sections, the lines of position are the curves that describe the possible location of the MS with respect to a single BS for each of the signal strength, AoA, and ToA methods. The curves are defined with respect to two BSs for TDoA. Each of the lines of position can be described mathematically with respect to the relative geometry of the BS and MS.

1.3.1 SIGNAL STRENGTH

Radiolocation using signal strength is a well known location method that uses a known mathematical model describing signal path loss with distance [63, 118, 161]. A path loss function characterizes the attenuation that a signal experiences as it travels over a distance away from a transmitter to a receiver. It is this path loss that helps define the cell sizes typical of cellular and PCS communications systems. Since a measurement of signal strength provides a distance, or range, estimate between the MS and BS, the MS must lie on a circle centered at the BS. By using multiple BSs, the location of the MS can be determined. This concept is illustrated in Fig. 3.4 where the range estimates r_1, r_2 and r_3 are determined from signal strength measurements.

For signal strength based location systems, the primary source of error is multipath fading and shadowing. Variations in the signal strength can be as great as 30-40 dB over distances on the order of a half wavelength. Signal strength averaging can help, but low mobility MSs may not be able to average out the effects of multipath fading and there will still be the variability due to shadow fading. Ott [118] showed with a path loss exponent of 3.5, a range estimate can be as much as 69% too large or 41% too small for an 8 dB shadow fade, which can be significant especially at large ranges, as illustrated in Fig. 3.5. A fuzzy logic algorithm was introduced in [70] to improve the accuracy. The errors due to shadow fading can also be combated by using pre-measured signal strength contours that are centered at the BSs [42, 159]. However, this approach assumes a constant physical topography (since shadows will change with the tree foliage, construction/destruction of structures, etc.) and requires that contours be mapped out for each BS.

In CDMA cellular systems, the MSs are power controlled to combat the near-far effect[2] [69, 95, 126]. TDMA cellular systems use power control to conserve battery power in the MSs. Therefore, for signal strength

[2]The *near-far effect* is a term used to describe the phenomenon that occurs when signals from some users are much stronger than other users' signals. As a result of the power imbalance, the stronger users' signals tend to mask out the signals of the weaker users making it difficult to detect and demodulate them.

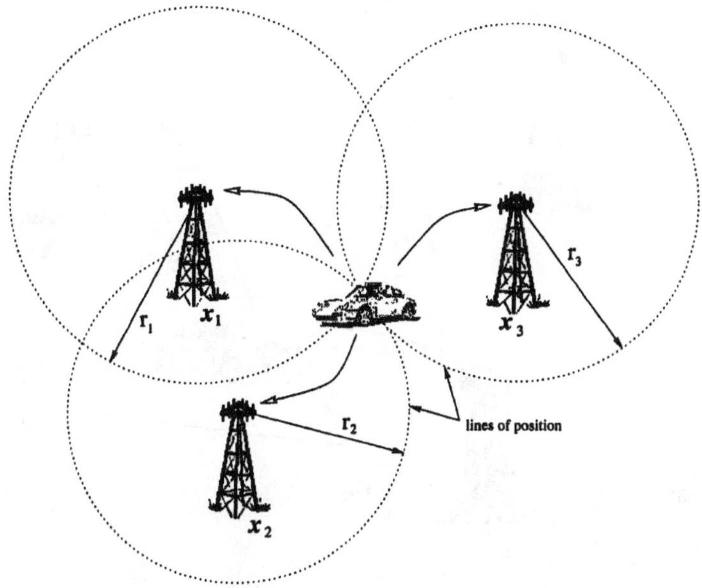

Figure 3.4. The geometry of the range-based signal strength location method. As will be shown, this model is also appropriate for ToA location. The lines of position are circles whose intersection provides the location of the MS.

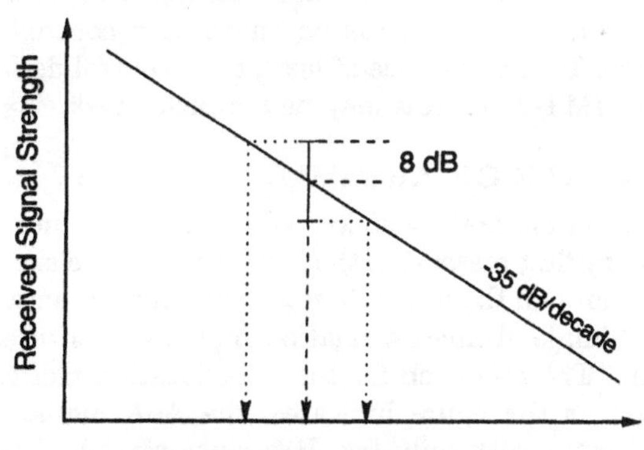

Figure 3.5. The error in range estimates caused by shadow variations in the received signal strength. The dashed arrow indicates the true range and the dotted arrows denote over- or under-estimated ranges due to shadowing.

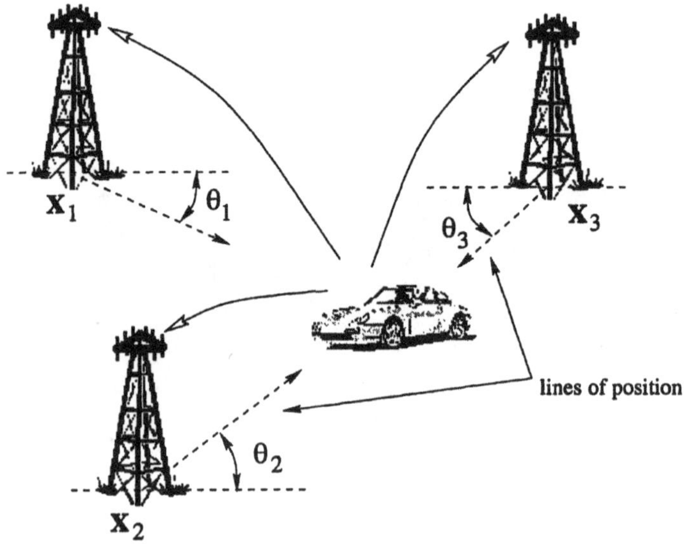

Figure 3.6. The angle of arrival method for MS location. The lines of position are straight lines whose intersection provides the location of the MS.

based location systems using reverse link signaling, it is necessary that the transmit power of the MSs be known and controlled with reasonable accuracy. In the presence of fast power control design, such as that proposed for IMT-2000, this may be a difficult task.

1.3.2 ANGLE OF ARRIVAL

AoA location methods, also known as direction finding, estimate the MS location by first measuring the arrival angles of a signal from a MS at several BSs through the use of directive antennas or antenna arrays [118, 139, 144]. Simple geometric relationships are then used to determine the location. The approach for the AoA location method is illustrated in Fig. 3.6. As the figure indicates, the AoA method can provide a location estimate with only two BSs since straight lines, defining the lines of position, are used for positioning.

The performance of an AoA location system is limited by the accuracy to which AoA estimates can be made. This is a characteristic of the hardware and estimation algorithm used. When such a location system is employed in a wireless network, performance becomes highly dependent upon the propagation environment since scattering near and around the MS and BS will effect the measured AoA [118]. The scatterers cause multiple signals to appear at the BS from the MS (and vice versa) which introduces error into the AoA estimates. In the absence of a LoS signal component, the antenna array will lock-on to a reflected signal that may not be coming from the direction of the MS. Even if a LoS component is present, multipath will still interfere with the angle measurement. The accuracy of the AoA method diminishes with increasing distance between the MS and BS due to the scattering environment and fundamental limitations of the devices used to measure the arrival angles. For instance, consider a scenario in which a measured AoA is in error by 3° at a certain BS. A MS that is located 200 m away from the BS will be 10 m away from the line of position, while a MS located 1000 m away will be 52 m from the line of position. This leads to larger error for the further MS since the location estimate is determined by the intersection of the lines of position.

In mobile radio environments with large cells (i.e., macrocells), scattering objects are primarily within a small distance of the MS. The scattering objects take the form of buildings, trees, and other objects that surround the MS. In contrast, the BSs are usually elevated well above the local terrain and thus are not surrounded by local scatterers [80, 121]. As a consequence, the signals arrive within a relatively narrow angular spread at the BSs. Jakes [80] and Gans [47] have modeled scattering in macrocells by assuming a ring of scatterers about the MS, with the BS situated well outside the ring.

In environments with smaller cells (i.e., microcells), the BS antennas may be placed below roof top level at lamp post height in order to confine the signal to a small area [61, 175]. As a result, the BSs will often be surrounded by local scatterers such that the signals arrive at the BSs with a large angular spread.

The scattering environment for macrocells and microcells is depicted in Fig. 2.3. For macrocells, the measured AoA will be in the general direction of the MS since it only receives signals over a narrow azimuth. For microcells, the measured AoA may be quite different from the true direction to the MS. Thus, while the AoA approach is useful for macrocells, it may be impractical for microcells unless provisions are made to guarantee LoS propagation.

1.3.3 TIME-BASED METHODS

The final class of radiolocation techniques are those based on estimating the ToAs of a signal transmitted by the MS and received at multiple BSs or the TDoAs of a signal received at multiple pairs of BSs

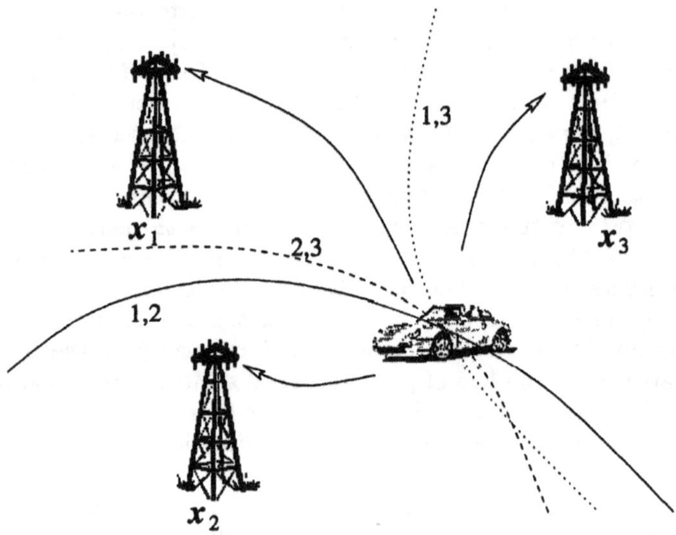

Figure 3.7. The time difference of arrival method for MS location. The lines of position are hyperbolas whose intersection provide the location of the MS.

[118, 141, 181]. In the ToA approach, the distance between a MS and BS is measured by finding the one way propagation time of a signal traveling between them. Geometrically, this provides a circle, centered at the BS, on which the MS must lie, as was the case for the signal strength location method shown in Fig. 3.4. By using at least three[3] BSs to resolve ambiguities arising from multiple crossings of the lines of position, the MS's position is given by the intersection of circles. In the TDoA approach, time differences of arrival are used. Since the hyperbola is a curve of constant time *difference* of arrival for two BSs, the lines of position are given by hyperbolas with foci at the BSs on which the MS must lie. The location of the MS is at the intersection of the hyperbolas shown in Fig. 3.7. The branches[4] of the hyperbolas chosen for the lines of position is determined from the sign of the TDoA measurement. If the sign is negative, we choose the hyperbola branch furthest from the BS with respect to which the TDoA measurement was made. If it is positive, the branch nearest the reference BS is chosen.

[3]Three BSs are required for location in two dimensions using the time-based methods.
[4]The reader may be familiar with hyperbola curves coming in pairs. The term *branch* used here corresponds to one of those hyperbola curve pairs.

The time-based radiolocation methods seem to have the greatest potential for providing high accuracy location estimates. Unlike the AoA methods, the accuracy of time-based methods does not degrade with increasing MS-BS distances, and it is not necessary to place large antenna arrays at the cell sites. Moreover, it is generally considered easier to obtain accurate time measurements as opposed to accurate angle measurements. The essential ingredient for the time-based approaches are high resolution timing measurements. However, it should be noted that LoS propagation conditions are still necessary to achieve high accuracy for the time-based methods.

Several methods have been proposed as means of forming time estimates in wireless systems. These include phase ranging, pulse ranging and spread spectrum techniques.

- **Phase Measurement.** In a phase ranging system, a low frequency tone is transmitted, and once received, applied to a phase detector from which the time of arrival is obtained [39, 181]. Once again, at least three BSs are required for a position fix. This system requires that the phase detectors be synchronized. Ambiguities arising from the unknown integer multiples of phase cycles can be resolved using frequency differencing or time of arrival information [41]. Location systems based on phase differencing (TDoA) are also a possibility (e.g., Decca, Omega). Significant error arises in phase ranging systems due to multipath propagation of the transmitted signal [37, 41, 141]. As a result, the composite signal will have a phase error relative to the direct signal which translates into a location error.

- **Pulse Transmission.** A pulse ranging system uses the transmission of a single pulse whose arrival time is measured at several receivers [41, 62, 181]. This technique can help reduce multipath effects since the first pulse to arrive can be considered to arrive via the most direct path. Unfortunately, the narrow pulses needed for high accuracy implies a large bandwidth. Time differences can also be formed using pulse transmission so that a TDoA location system can be implemented. A commercial application of a pulse location system is Loran [39, 135].

- **Spread Spectrum Techniques.** With spread spectrum signaling, the ToAs or TDoAs can be determined through the use of correlation techniques. Due to the nature of the spread spectrum signals, the effects of multipath can be reduced. This is a result of the periodic autocorrelation function of the spreading sequences which is typically small for delays greater than a chip period [67, 207]. Applications of spread spectrum to ranging have previously been investigated (e.g., [16, 17, 55]) and is the principle behind GPS [36].

2. ADVANTAGES/DISADVANTAGES OF LOCATION METHODS

Each of the methods described in the previous section have their advantages and disadvantages. Dead-reckoning does not require a large infrastructure of devices to locate a MS. Rather, it relies on specialized equipment for self-positioning of a MS. This is a disadvantage in wireless communication systems because additional equipment must be purchased by the wireless subscriber and a method must be developed whereby the calculated location information can be communicated to the network. This would most certainly require modification to the handset in the MS. Location by proximity detection enables coarse positioning, but suffers the disadvantage of large infrastructure costs to outfit an area with sensors. As mentioned above, sectors of cells may be used for proximity detection; however, the accuracy that is required of E-911 and other services cannot be met using such a method unless other devices are added to the area to improve location accuracy. This again adds costs for both the planning of the system (placement of sensors for coverage, etc.) and the deployment of the sensors. Thus, dead-reckoning and proximity detection do not provide viable means of position location in wireless communication systems.

Radiolocation, on the other hand, finds a natural fit in wireless communication systems since it is based on radio signaling. Since the E-911 requirements cover existing as well as future MS handsets, methods of location are necessary which require no modification to the handsets. Consequently, remote-positioning is the method of choice. In this way, the pre-existing infrastructure of transmitters and receivers can be used for location. Since cellular and PCS systems are wireless, self-positioning is also a possibility. However, self-positioning requires that the MS handset have "built-in" location technology. For future generation cellular and PCS systems, location services can be built directly into emerging and future standards.

In the next section and in the later chapters of the book, only the impairments to accurate location that are experienced by radiolocation methods are discussed due to the propagation channel.

3. SOURCES OF LOCATION ERROR

Error can be introduced to the location process in a variety of ways. Obviously, the equipment that is used to measure location parameters such as AoAs, ToAs and TDoAs will place limitations on the accuracy that can be achieved by a location system. Separate from the location equipment is the surrounding environment which also plays a key role in the achieved location accuracy. The analyses and impairments discussed throughout this book assume that the equipment used to measure location parameters is ideal, i.e., no errors arise due to the operation of the equipment. Thus, antenna array or correlator measurements introduce no error into the system themselves. The only error that is considered is from the propagation medium over which signals must pass before they are measured by the BS equipment (assuming reverse link location). In this regard, sources of location error in wireless communication systems include multipath propagation and non-line-of-sight propagation [16, 118, 154]. For CDMA, interference from other users also contributes to the error because of the near-far effect [16]. Steps must be taken to mitigate these impairments to improve the location accuracy. In the following, we discuss the effect that each of these impairments has on the ability to accurately locate a MS in a CDMA cellular or PCS environment. Although the effects of theses impairments are discussed for all radiolocation systems, particular emphasis is placed on the time-based methods, as these will be the focus of the later chapters.

3.1 MULTIPATH PROPAGATION

Multipath propagation is described by the reception of multiple signals at a MS or BS. These signals can combine constructively (larger signal) or destructively (smaller signal) and result in the phenomenon known as *fading*. The multiple signals that are received make it difficult to accurately determine the signal strength, AoAs, ToAs or TDoAs in wireless systems. As mentioned in Section 1.3, the high variability of the received signal strength in mobile radio systems makes it difficult to use path loss models for location. Although it is possible to estimate the multiple AoAs of multipath at an antenna array using appropriate processing techniques, the AoA corresponding to the true AoA must be determined.

For time-based location systems, multipath propagation causes errors in the timing estimates even when there is a LoS path between the MS and BS. Conventional propagation delay and differential time delay estimators, which are usually based on correlation techniques, can be used for CDMA location and are influenced by the presence of multipath, especially when the reflected rays arrive within a chip period of the first arriving ray. When the first ray arrives with less power than later arriving rays, conventional delay estimators will detect a delay in the vicinity of these later arriving rays. Several authors have studied the

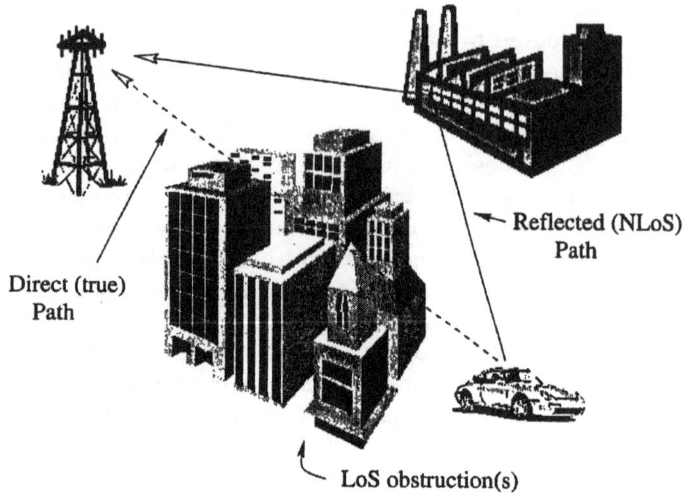

Figure 3.8. Propagation of signals when LoS path does not exist. The first path to arrive at the BS is a reflection that travels a greater distance and arrives at the BS from a different direction than the true path.

effects of multipath on conventional code acquisition [72, 162] and code tracking techniques [151]. The multipath component is seen to bias the tracking of the DLL. Several methods have been developed to mitigate the effects of multipath on delay estimation, including a high resolution frequency estimator [87] and a least-mean-squares (LMS) technique [160]. Super-resolution techniques such as the MUSIC, Root-MUSIC and the TLS-ESPRIT algorithm have been utilized to detect multipath components that conventional detectors are unable to detect [35, 120, 145]. Delay estimators based on the EKF [46, 76, 77] as well as the conventional DLL [152, 177, 185] have also been developed for multipath corrupted signals.

3.2 NLOS PROPAGATION

Non-line-of-sight (NLoS) propagation describes the scenario where the direct, or LoS, path between the MS and BS is blocked by some structure (e.g., buildings, mountains). With NLoS propagation, the signal arriving at the BS from the MS is reflected or diffracted and takes a path that is longer than the direct path, as illustrated in Fig. 3.8. Note that the NLoS problem is not only an impairment in CDMA systems, but equally affects all types of cellular systems (TDMA, FDMA). For signal strength location systems, several propagation models exist for both macrocells and microcells [134, 175]. For AoA location systems, large errors can be incurred especially in microcells where local scatterers surround the BS. According to Fig. 3.8, if the reflected path is measured for AoA location, the line of position from that BS will lie far from the true position of the MS.

For time-based location systems, the NLoS path can be much longer than the LoS path resulting in ToA or TDoA measurements that are in error by the extra path length. If the reflected paths received at two BSs from an MS are of the same length, then the TDoA measurements will not be affected. The typical ranging error introduced by NLoS propagation has been measured in the GSM system which indicates that NLoS error can average between 500-700 meters [154]. NLoS propagation will bias the ToA or TDoA measurements even when high resolution timing techniques are employed and even if there is no multipath interference. Therefore, it is important to find methods to mitigate the NLoS error. One such method is to distinguish LoS and NLoS BSs by measuring the standard deviation of the ToA measurements taken over a period of time as the MS moves about the cell [197]. The standard deviation of the range measurements is much higher for NLoS propagation than LoS propagation [154]. With knowledge of the range error statistics, the range measurements made over a period of time and corrupted by NLoS error can be adjusted to values near their correct LoS values. A second approach is to use an algorithmic technique which attempts to reduce the contribution of NLoS corrupted measurements to the final location estimate [16]. This approach, however, requires a means of determining which BSs are NLoS as discussed in Chapter 5. These two methods will be discussed in further detail in Chapter 7.

3.3 MULTIPLE-ACCESS INTERFERENCE

All cellular systems suffer from co-channel interference. In cellular CDMA, users share the same frequency band with different spreading codes. Hence, their reverse link transmissions interfere with one another resulting in the near-far effect, making it difficult to recover the weaker users [102, 126, 190]. With forward link based location, the interference arises from other BSs.

Power control schemes, which attempt to ensure that each user's signal is received with equal power at the BS [69, 95] can be used to combat the near-far effect. For location in a CDMA cellular system, the near-far effect remains a factor even when power control schemes are used since the interference will affect the ability of a conventional receiver at other BSs to estimate the ToA or TDoA information. For example, consider reverse link location in the macrocell deployment shown in Fig. 3.9,

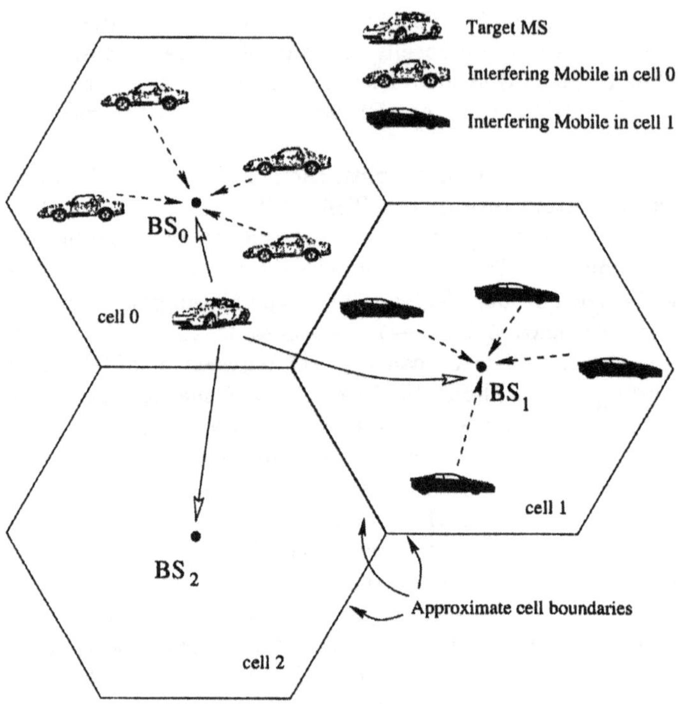

Figure 3.9. CDMA multiuser interference scenario for macrocells.

where the target MS whose location is desired is being served by BS_0. To determine its location, BS_0, BS_1 and BS_2 are used. If power control is used, then the signals from all MSs served by BS_0 (those shown in cell 0) will arrive at BS_0 with approximately the same power. The same is true for the MSs being served by BS_1 and BS_2 (not shown). To derive the location estimate, BS_1 and BS_2 must detect the signal being transmitted by the target MS and derive ToA or TDoA information from it. However, the signal from the target MS may experience severe MAI from MSs being served by BS_1 and BS_2, since the target MS is not power controlled by those BSs. This is particularly the case when the target MS is close to its serving BS. The closer the MS is to its serving BS the smaller its transmit power must be compared to MSs further out from the BS. Since the target MS will be transmitting smaller power, it will be more difficult to detect its signal at the other BSs due to the other user interference at those BSs. This phenomenon has been referred to as the problem of "hearability" [135]. For future generation wireless systems, it may be possible to perform forward link location, where the MS calculates its own position (or at least the location parameters) by listening to the pilot channels of nearby BSs. In IS-95, the MS listens to the pilot channels to aid in the handoff process, so the foundation is present for further development.

As the next chapter will discuss, the time-based methods typically use correlation techniques to derive the ToA and TDoA estimates for use in a location algorithm. In spread spectrum systems, such as CDMA, the conventional means of obtaining ToA estimates is through coarse or fine propagation delay acquisition while TDoA estimates are obtained through cross-correlation techniques. It has been shown that MAI greatly affects the coarse code acquisition of spread spectrum signals [29, 104]. Likewise, the effects of MAI on code tracking techniques, such as the conventional DLL, have been shown to be quite drastic [15]. This has important implications for location systems that use conventional techniques to provide ToA or TDoA estimates. However, it may be possible, for instance in E-911 situations, for the MS to power up to maximum level and, therefore, mitigate the near-far effect. A further possibility is to take advantage of soft handoffs. A soft handoff occurs when a MS is in the region between two or more cells, and communicates with the other BSs to achieve a macrodiversity gain [191]. The soft handoff allows a smoother handoff with less chance for a dropped call. However, for soft handoffs to be useful for a time based location system, at least three BSs must be involved in the soft handoff, an unlikely situation [51].

As a consequence of the research into near-far resistant CDMA multiuser detection, several authors have investigated near-far resistant delay estimators. With the interference problem that was just discussed, these provide the means of forming accurate ToA estimates for location in the presence of MAI. Many of the near-far resistant delay estimators are based on subspace techniques such as MUSIC [8, 81, 174]. Other estimators have been produced in conjunction with multiuser detectors [78, 133] and interference cancellation techniques [89, 205]. Further details of these methods will be discussed in Chapter 6.

4. SUMMARY

This chapter has reviewed several techniques for locating MSs. The method of radiolocation fits nicely in with the radio framework of wireless communications networks. Several impairments to accurate radiolocation that arise in mobile radio channels were discussed. These impairments consisted of multipath propagation, multiple-access interference and NLoS propagation. Each of these must be accounted for in order to achieve accurate location in wireless systems. The latter two will be addressed in more detail in later chapters. We did not cover sources of error that arise from hardware limitations.

Chapter 4

ALGORITHMS FOR RADIOLOCATION

The ability to locate a MS is dependent on the acquisition of accurate information that can be used for positioning. Location parameters, such as AoAs, ToAs and TDoAs, must first be measured accurately before an algorithm can be utilized to calculate an estimate of the MS's position. Hence, we see that the process of calculating the location estimate can be viewed as consisting of two parts. The first part provides estimates of location parameters and the second part is the algorithm that uses the parameters to determine the MS's position. In this chapter, various methods for estimating the location parameters are discussed along with the algorithms that use those parameters for location.

1. ESTIMATION OF LOCATION PARAMETERS

Location parameters are determined from the measurement of certain signal characteristics. The measurements are provided by the hardware and/or software of the receiver and is thus subject to error. In the sections that follow, the only sources of error that are considered come from the propagation channel, not from the hardware.

1.1 SIGNAL STRENGTH ESTIMATION

Before the empirical models describing path loss attenuation with distance between the MS and BS can be employed, an estimate of the received power at the BS must be made. Once the estimate is obtained, the path loss model can be used to compute the range to the MS, assuming its transmit power is known. A mobile radio envelope is composed of fast (multipath) fading superimposed on a slow (shadow) faded signal.

The local mean power of the signal can be obtained by averaging out the fading. Two approaches, one analog and the other discrete, are available for computing the local mean power, both of which require proper choice of window length for averaging [4, 92]. The received signal strength at some position l can be described by the multiplicative model [92]

$$r^2(l) = r_p(l) \cdot m_p(l) \tag{4.1}$$

where $r^2(l)$ is the squared envelope, $r_p(l)$ is either Rayleigh or Rician depending on whether a specular, or LoS, path exists, and $m_p(l)$ is a log-normal random variable describing the shadow fading. If the local mean is constant, then $m_p(l) = m_p$. Assuming ergodicity, an integral spatial average of $r^2(l)$ can be used to estimate the local mean m_p:

$$\hat{m}_p = \frac{1}{2L}\int_{x-L}^{x+L} r^2(l)dl \tag{4.2}$$

$$= \frac{m_p}{2L}\int_{x-L}^{x+L} r_p(l)dl \tag{4.3}$$

where the second equality follows from the assumption that the mean is constant over the interval $(x-L, x+L)$. If $2L$ is too large, the estimated local mean will not describe the log-normal shadowing. If it is too short, the estimate will still contain multipath fading information.

More practical signal strength estimators use samples of the signal strength rather than analog averaging. Consider the sampled signal

$$r^2(i) \triangleq r^2(iD) \tag{4.4}$$

where D is the spatial sampling period and i is an integer. Then the sample average

$$\hat{m}_p = \frac{1}{N_s}\sum_{i=0}^{N_s-1} r^2(i) \tag{4.5}$$

can be used as an estimate of the local mean power for N_s samples.

The accuracy of the estimates from both approaches can be obtained by determining the variance of (4.3) and (4.5). These values have been derived for both Rayleigh [92] and Rician [4] fading environments.

The ranges obtained using the above methods are still too rough for accurate location estimation. To improve the accuracy, a fuzzy logic technique was developed that further manipulates the raw range data in order to achieve an acceptable level of accuracy [161]. The approach makes use of membership functions to provide a measure of the reliability of the estimated range. Simulations showed that accuracy could be greatly improved, although not the degree required by the wireless E-911 regulations.

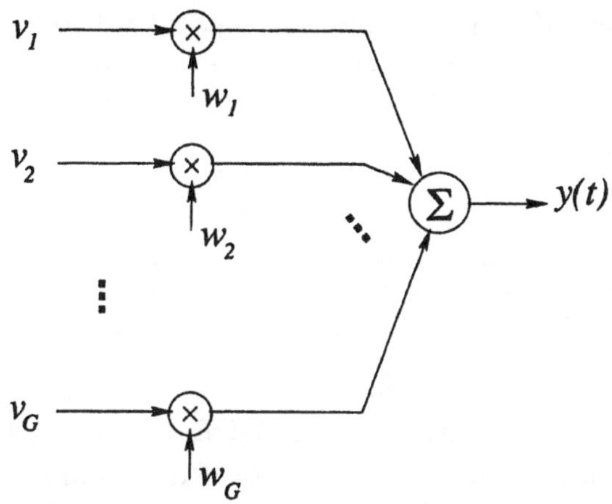

Figure 4.1. Antenna array for beamforming.

1.2 ANGLE OF ARRIVAL ESTIMATORS

Several methods have been developed for estimation of the AoAs at a receiver [52]. Although the remaining chapters of this book focus on the use of time-based location methods for wireless location, techniques for AoA estimation will be briefly reviewed for the interested reader.

The AoAs are typically estimated through the use of antenna arrays at the BSs. An antenna array is composed of a number of antenna elements whose signals are often combined to produce a beam in a desired direction. In mobile radio systems, the antenna arrays are typically located only at the BSs, since it is difficult to place an array in a MS handset. Thus, AoA estimation is generally used for remote-positioning.

Consider the antenna array shown in Fig. 4.1 in which samples from Q source signals are sampled at G antenna array elements. A single snapshot of the signals received by the array can be expressed compactly as

$$\boldsymbol{v}(t) = \boldsymbol{A}(\boldsymbol{\Phi})\boldsymbol{s}(t) + \boldsymbol{n}(t) \tag{4.6}$$

where $\boldsymbol{v}(t) = [v_1(t), \ldots, v_G(t)]^T$, $\boldsymbol{A}(\boldsymbol{\Phi})$ is the $G \times Q$ matrix of steering vectors defined by

$$\boldsymbol{A}(\boldsymbol{\Phi}) = [\boldsymbol{a}(\phi_1), \ldots, \boldsymbol{a}(\phi_Q)] \quad , \tag{4.7}$$

and $\boldsymbol{s}(t)$ is the $Q \times 1$ vector of source signals given by

$$\boldsymbol{s}(t) = [s_1(t), \ldots, s_Q(t)]^T \quad . \tag{4.8}$$

The notation $[\cdot]^T$ and $[\cdot]^H$, which will appear later, denote the transpose and the conjugate transpose of a matrix, respectively. For N_s total snapshots, the sampled signals at the G antenna elements can be expressed as

$$\boldsymbol{V} = \boldsymbol{A}(\boldsymbol{\Phi})\boldsymbol{S} + \boldsymbol{N} \tag{4.9}$$

where $\boldsymbol{V} = [\boldsymbol{v}(0), \ldots, \boldsymbol{v}(N_s - 1)]$ and $\boldsymbol{N} = [\boldsymbol{n}(0), \ldots, \boldsymbol{n}(N_s - 1)]$ are $G \times N_s$ matrices, and $\boldsymbol{S} = [\boldsymbol{s}(0), \ldots, \boldsymbol{s}(N_s - 1)]$ is a $Q \times N_s$ matrix.

The goal is to estimate the AoAs, $\phi_1, \ldots, \phi_Q$, from the N_s snapshots in the array, $\boldsymbol{V}$. Methods to accomplish this task are briefly discussed below.

1.2.1 MAXIMUM LIKELIHOOD (ML) ESTIMATORS

One of the first methods to be investigated for AoA estimation was the ML technique. The development shown here follows that of [208]. A few assumptions are necessary before the development of the ML estimator can begin. First, the number of signals is assumed to be known and is smaller than the number of sensors, i.e., $Q < G$. Second, the set of steering vectors is linearly independent. Third, the noise is assumed to be a stationary and ergodic complex-valued Gaussian process with zero mean and variance matrix $\sigma_n^2 \boldsymbol{I}$, where σ_n^2 is an unknown scalar, and $\boldsymbol{I}$ is the identity matrix. Finally, the noise samples are assumed to be statistically independent.

With these assumptions, the log-likelihood function, derived from the joint density function of the received sampled signals $\boldsymbol{V}$, is given by [208]

$$\Lambda = -N_s G \log \sigma^2 - \frac{1}{\sigma_n^2} \sum_{i=0}^{N_s-1} |\boldsymbol{v}(i) - \boldsymbol{A}(\boldsymbol{\Phi})\boldsymbol{s}(i)|^2 \tag{4.10}$$

where any constant terms have been ignored. The ML estimator is computed by maximizing the log-likelihood function of (4.10) with respect to the unknown parameters. Because the logarithm is a monotonic function, the maximization problem can be replaced by an equivalent minimization problem:

$$\min_{\boldsymbol{\Phi},\boldsymbol{S}} \left\{ \sum_{i=0}^{N_s-1} |\boldsymbol{v}(i) - \boldsymbol{A}(\boldsymbol{\Phi})\boldsymbol{s}(i)|^2 \right\} \tag{4.11}$$

which happens to be the least squares (LS) criterion for the estimation problem. Holding $\boldsymbol{\Phi}$ fixed and minimizing with respect to $\boldsymbol{S}$ yields the well known LS solution

$$\hat{\boldsymbol{s}}(i) = \left[\boldsymbol{A}^H(\boldsymbol{\Phi})\boldsymbol{A}(\boldsymbol{\Phi})\right]^{-1} \boldsymbol{A}^H(\boldsymbol{\Phi})\boldsymbol{v}(i) \ . \tag{4.12}$$

Substituting (4.12) into (4.11) gives the following minimization problem:

$$\min_{\Phi}\left\{\sum_{i=0}^{N_s-1}\left|v(i) - P_{A(\Phi)}v(i)\right|^2\right\} \tag{4.13}$$

where $P_{A(\Phi)}$ is the project operator onto the space spanned by the columns of $A(\Phi)$,

$$P_{A(\Phi)} = A(\Phi)\left[A^H(\Phi)A(\Phi)\right]^{-1}A(\Phi)\ . \tag{4.14}$$

Thus, the ML estimate of the AoA vector Φ is obtained by maximizing the log-likelihood function

$$\begin{aligned}\Lambda_{\Phi} &= \sum_{i=0}^{N_s-1}\left|P_{A(\Phi)}v(i)\right|^2 &(4.15)\\ &= \text{trace}\left[P_{A(\Phi)}\hat{R}\right] &(4.16)\end{aligned}$$

where $\hat{R}$ is the sample covariance matrix

$$\hat{R} = \frac{1}{G}\sum_{i=0}^{N_s-1} v(i)v^H(i)\ . \tag{4.17}$$

Maximization of the log-likelihood function in (4.16) is a nonlinear, multidimensional maximization problem, and as such is computationally expensive. Consequently, other methods have been developed to reduce the complexity. One such technique, the *Alternating Projection* method, uses an iterative technique to transform the multidimensional, nonlinear problem to a sequence of simpler one-dimensional problems [208]. Details of other ML methods can be found in [52].

Due to the complexity of the ML approach to AoA estimation, several suboptimal approaches have been developed. A few of these are discussed next.

1.2.2 MINIMUM VARIANCE METHOD

The minimum variance approach seeks to minimize the average power of the beamformer output while a distortionless response is maintained along the direction of a target signal of interest [64]. Hence, it is also known as the Minimum Variance Distortionless Response (MVDR) estimator. As shown in Fig. 4.1, the output of the antenna array at some time instant i is given by

$$y(i) = w^H v(i) \tag{4.18}$$

where $\boldsymbol{w}$ is the $G \times 1$ weight vector of the beamformer. The output power of the beamformer is given by

$$P_{bf} = \mathrm{E}\left[|y(i)|^2\right] \tag{4.19a}$$

$$= \boldsymbol{w}^H \boldsymbol{R}_{v_i,v_i} \boldsymbol{w} \tag{4.19b}$$

where the second equality follows from substituting (4.18) into (4.19a) and defining $\boldsymbol{R}_{v_i,v_i} = \mathrm{E}\left[\boldsymbol{v}(i)\boldsymbol{v}^H(i)\right]$ as the autocorrelation of the array inputs.

The minimum variance method is based on the following minimization problem:

$$\min_{\boldsymbol{w}} \boldsymbol{w}^H \boldsymbol{R}_{v_i,v_i} \boldsymbol{w} \qquad \text{subject to} \qquad \boldsymbol{w}^H \boldsymbol{a}(\phi_0) = 1 \tag{4.20}$$

where ϕ_0 is the angle of interest. To solve this constrained minimization problem, the method of Lagrange multipliers is used to form the MVDR solution for the weight vectors, $\hat{\boldsymbol{w}}$. Using these weight vectors, the MVDR spatial spectrum [64]

$$P_{\mathrm{MVDR}} = \frac{1}{\boldsymbol{a}^H(\phi)\boldsymbol{R}_{v_i,v_i}^{-1}\boldsymbol{a}(\phi)} \tag{4.21}$$

provides the output power as a function of the angle of arrival, the peaks of which indicate the estimated AoAs.

1.2.3 SUBSPACE TECHNIQUES

In subspace AoA estimation methods, the signals measured at each element of a G-element array are seen as defining a G-dimensional space. When there are more array elements than signals present ($Q < G$), the signal component of $\boldsymbol{v}$ is confined to (at most) a Q-dimensional subspace, referred to as the *signal subspace*. Schmidt [148] used this geometric interpretation to develop the MUSIC (MUltiple SIgnal Classification) algorithm, the geometric concepts of which form the basis for a broader class of subspace-based algorithms [125]. The MUSIC algorithm is based on exploiting the eigenstructure of the input covariance matrix. In terms of the input data model in (4.6), the input covariance matrix is expressed as

$$\boldsymbol{R}_{v,v} = \mathrm{E}\left[\boldsymbol{v}\boldsymbol{v}^H\right] \tag{4.22}$$

$$= \boldsymbol{A}(\boldsymbol{\Phi})\mathrm{E}\left[\boldsymbol{s}\boldsymbol{s}^H\right]\boldsymbol{A}^H(\boldsymbol{\Phi}) + \mathrm{E}\left[\boldsymbol{n}\boldsymbol{n}^H\right] \tag{4.23}$$

$$= \boldsymbol{A}(\boldsymbol{\Phi})\boldsymbol{R}_{ss}\boldsymbol{A}^H(\boldsymbol{\Phi}) + \sigma_n^2 \boldsymbol{I} \tag{4.24}$$

where $\boldsymbol{R}_{ss}$ is the signal correlation matrix and $\boldsymbol{A}(\boldsymbol{\Phi})\boldsymbol{R}_{ss}\boldsymbol{A}^H(\boldsymbol{\Phi})$ is a rank Q matrix (assuming none of the signals are coherent). An eigendecomposition of $\boldsymbol{R}_{v,v}$ provides its eigenvalues $\{\lambda_i\}_{i=1}^G$ and eigenvectors $\{\boldsymbol{e}_i\}_{i=1}^G$. Since $\boldsymbol{R}_{v,v}$ is composed of a rank Q matrix plus a scaled identity, then the $G-Q$ smaller eigenvalues are repeated and are equal to σ_n^2. Assuming that the eigenvalues are ordered in decreasing magnitude, $\lambda_{Q+1} = \cdots = \lambda_G = \sigma_n^2$. If we define $\boldsymbol{E}_S = [\boldsymbol{e}_1, \ldots, \boldsymbol{e}_Q]$ and $\boldsymbol{E}_N = [\boldsymbol{e}_{Q+1}, \ldots, \boldsymbol{e}_G]$, it is easy to see that the span of the columns of $\boldsymbol{E}_S$ span the signal subspace [125]. Its orthogonal complement, $\boldsymbol{E}_N$ is referred to as the noise subspace.

In practice, the ideal covariance matrix $\boldsymbol{R}_{v,v}$ is rarely available and must be approximated based from the average of N_s snapshots. The noise subspace is then obtained by selecting the eigenvectors corresponding to the $G-Q$ smallest eigenvalues of the estimated covariance matrix. By noting that the steering vectors corresponding to the signal components are orthogonal to the noise subspace vectors, the AoAs can be estimated by locating the peaks of a MUSIC spatial spectrum given by [52]

$$P_{\text{MUSIC}}(\phi) = \frac{1}{\boldsymbol{a}^H(\phi)\boldsymbol{E}_N\boldsymbol{E}_N^H\boldsymbol{a}(\phi)} . \tag{4.25}$$

The orthogonality between $\boldsymbol{a}(\phi)$ and $\boldsymbol{E}_N$ minimizes the denominator and causes the peaks in the MUSIC spectrum.

Variants of the MUSIC algorithm have been developed to improve its resolution and decrease its computational complexity including Root-MUSIC [6] and Cyclic MUSIC [146]. Improved subspace-based AoA estimation techniques include the ESPRIT algorithm [124] and its variants, and a minimum-norm approach [88]. Further details of subspace based estimation techniques can be found in [52] and [125].

1.2.4 CONVENTIONAL METHODS

A very simple approach for estimating the AoAs at an antenna array is to search for the angles which maximize the output power of the classical beamformer of Fig. 4.1. As in the minimum variance estimator, the output power is defined by P_{bf} in (4.19b). For a signal arriving with an angle ϕ_0 at the array, it can be shown that the power is maximized when the weights of the array, $\boldsymbol{w}$, equal the steering vector associated with the angle ϕ_0, $\boldsymbol{a}(\phi_0)$. Consequently, the estimated AoAs are given by the peaks of the power function, P_{bf}. Unfortunately, this technique has poor resolution as compared to other AoA estimation algorithms.

Another method discussed in [144] uses multibeam antennas to estimate the AoAs. Given the antenna patterns of a three-beam antenna, signal strength measurements at the antenna are matched against the three antenna patterns to provide an estimate of the AoA of an incoming signal. However, the accuracy of this method in a multipath environment remains unclear.

1.2.5 DISCUSSION OF AOA ESTIMATORS

In general, the ML method provides superior performance to the other AoA estimation techniques and can perform well when the input signals are correlated. However, the complexity of the ML estimator make it less popular than the subspace techniques which can provide good resolution of closely spaced AoAs. However, the subspace methods fail when the input signals to the antenna array are correlated, as in a multipath fading environment. This problem arises because the basis of the MUSIC algorithm requires that the signals be uncorrelated so that the source covariance $\boldsymbol{R}_{ss}$ is full rank. A technique known as *spatial smoothing* can be used to improve the performance of MUSIC in correlated signal environments. The interested reader is referred to [150].

1.3 TOA AND TDOA ESTIMATORS

The time based methods rely on accurate estimates of the ToAs or TDoAs from the signals received at several BSs or at the MS. Several approaches have been developed for estimation of these parameters from the received signals, a few of which are discussed below.

1.3.1 TOA ESTIMATION

Since we are concerned with location in CDMA networks, we first examine methods of estimating propagation delays using spread spectrum signaling. Delay estimation in spread spectrum systems usually consists of two parts: code acquisition and code tracking [67, 155, 207]. Code acquisition refers to the initial synchronization of the received spreading code and the receiver's local spreading code. Code tracking attempts to maintain synchronization after initial acquisition. Both of these methods can be used for the location. In fact, previous location studies in CDMA systems have used code acquisition to obtain ToA estimates [17, 55, 65]. Additionally, the performance of ToA location using estimates from code tracking loops has been studied in [16].

In the following, methods of code acquisition and tracking are discussed that are designed to operate in additive white Gaussian noise (AWGN) and multipath channels. Discussion of techniques for acquisition and tracking in multiple-access channels is reserved for Chapter 6.

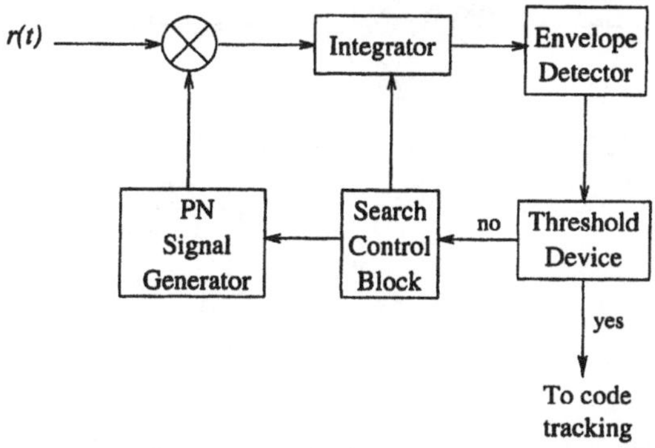

Figure 4.2. The sliding correlator for initial code acquisition.

Conventional code acquisition. The process of code acquisition typically determines the time delay estimate to within a "chip" time through the use of a sliding correlator, matched filter, or sequential acquisition circuit [67, 155, 207]. These methods seek to solve the maximization problem

$$\max_{\tau} \int_{0}^{LT_c} c(t+\tau)c(t)\, dt \tag{4.26}$$

by searching over the possible unknown code phases, or "cells", where L is the length of the code sequence over which the search is performed.

The sliding correlator shown in Fig. 4.2 is a simple method of performing initial code acquisition. It operates by performing a full or partial correlation between the incoming code with the local code at the receiver. If a received code is not detected, the local code used in the correlator is advanced by a fixed amount, usually one-half chip, and correlation is repeated. If the received code is detected, the received signal is handed over to the code tracking unit, discussed below. The circuit shown in Fig. 4.2 is often referred to as a single dwell-time acquisition system which suffers from its long acquisition time. A generalization of this system produces the double dwell-time system [67], or multiple dwell-time systems [207]. The double dwell time system has two correlators, one to search code phases quickly and the other to provide a further indication of whether the correct code phase has been found.

An efficient method of initial synchronization is to use a matched filter on the received signal. A digital implementation of a matched filter is illustrated in Fig. 4.3. Unlike the sliding correlator, the matched filter can be sampled q times per chip allowing it to provide sub-chip resolution at the expense of faster sampling hardware only. To achieve

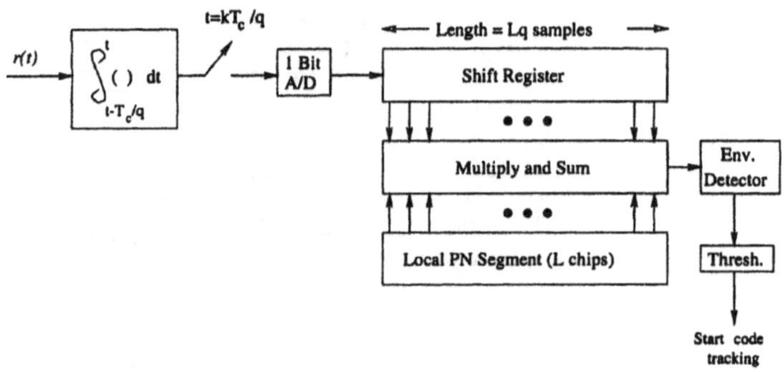

Figure 4.3. The matched filter for initial code acquisition. L is the length of code for correlation.

better than one-half chip resolution for the sliding correlator requires additional acquisition time to search additional cells [17].

The two code acquisition schemes above are known as *serial search* schemes in which the unknown code phase of the incoming signal is searched for in a serial manner. To improve acquisition times, *parallel search* schemes examine multiple code phases at once [26, 140].

Initial code synchronization can also be obtained by estimating the received spreading code [192]. Methods using this approach use *maximal length* spreading codes that can be generated using shift registers. One form of the shift register implementation always contains s symbols of the code sequence [207]. Thus, if s symbols of the spreading sequence can be estimated from the received waveform, these symbols can be loaded into the shift register generator to synchronize the system. This technique is referred to as rapid acquisition by sequential estimation (RASE) [192].

Fourier-based code acquisition. An algorithm in [86] uses phase spectrum data to exploit the periodicity between the phase spectra of the received sequence and the local reference. The discrete Fourier transforms (DFTs) of the received and local sequences differ by a complex-exponential factor related to the code offset between those sequences. Thus, the phase difference between the signals is

$$\theta(k) = \arg(L(k)) - \arg(R(k)) = \frac{2\pi\tau k}{N_D} \tag{4.27}$$

where $L(k)$ and $R(k)$ are the DFTs of the local and received signals, τ is the code offset, and N_D is the length of the DFT. With phase wrapping, $\theta(k)$ is a sawtooth waveform. The slope of the waveform then provides an estimate of the delay, given N_D, but $\theta(k)$ is noisy due to the noise in the received signal. An alternative is to take the DFT of $\theta(k)$ and determine the fundamental frequency of the sawtooth waveform. The fundamental frequency of the resulting DFT will indicate the magnitude of τ. The sign of the imaginary part of the DFT will indicate the sign of τ.

Subspace-based code acquisition. Subspace techniques have been developed for code acquisition in multipath fading channels since the conventional methods perform poorly in such environments. The super resolution offered by these algorithms provides a means of resolving closely spaced multipath components which are unable to be resolved using conventional correlation techniques. As in the case of AoA estimation, MUSIC has been used for time delay estimation in multipath faded channels [120]. Consider the received signal given in (2.21) without data modulation ($d_{1,m} = 1$) and with a single user ($K = 1$) that is passed through a filter matched to the user's spreading code, $a(t)$, yielding the output

$$y(t) = \int r(t)a(t)dt = \sum_{i=1}^{M} \beta_i R_a(t - \tau_i) + n_a(t) \tag{4.28}$$

where $R_a(t)$ is the autocorrelation function of the spreading sequence and $n_a(t)$ is the cross-correlation of the spreading code and the noise. Taking N_s samples of the matched filter output every T_s seconds, we obtain

$$\boldsymbol{y} = \boldsymbol{R_a B} + \boldsymbol{n_a} \tag{4.29}$$

where

$$\begin{aligned} \boldsymbol{R_a} &= \begin{bmatrix} R_a(-\tau_1) & \cdots & R_a(-\tau_M) \\ R_a(T_s - \tau_1) & \ddots & \\ \vdots & & \\ R_a((N_s-1)T_s - \tau_1) & \cdots & R_a((N_s-1)T_s - \tau_M) \end{bmatrix} \\ &= [\boldsymbol{c}_1 \ \boldsymbol{c}_2 \ \cdots \ \boldsymbol{c}_M] \end{aligned} \tag{4.30}$$

with $\boldsymbol{c}_i$ representing the ith column of $\boldsymbol{R_a}$,

$$\boldsymbol{B} = \begin{bmatrix} \beta_1(0) & \beta_1(1) & \cdots & \beta_1(N_s-1) \\ \beta_2(0) & \ddots & & \\ \vdots & & & \\ \beta_M(0) & \beta_M(1) & \cdots & \beta_M(N_s-1) \end{bmatrix} \tag{4.31}$$

and $\boldsymbol{n}_a = [n_a(0), \cdots, n_a(N_s - 1)]^T$. The covariance matrix $\mathrm{E}[\boldsymbol{y}\boldsymbol{y}^H]$ can be shown to be [120]

$$\boldsymbol{R}_y = \boldsymbol{R}_a \boldsymbol{\Gamma}_B \boldsymbol{R}_a^H + \boldsymbol{\Gamma}_n \tag{4.32}$$

where $\boldsymbol{\Gamma}_B = \mathrm{E}[\boldsymbol{B}\boldsymbol{B}^H]$ and $\boldsymbol{\Gamma}_n = \mathrm{E}[\boldsymbol{n}_a \boldsymbol{n}_a^H]$. Using the MUSIC algorithm and assuming that $\boldsymbol{B}$ is full rank (i.e., channel attenuations are uncorrelated), the delays τ_k can be estimated from

$$\max_{\tau} \left\{ \frac{\boldsymbol{c}^H(\tau)\boldsymbol{c}(\tau)}{\sum_{i=M+1}^{Q} |\boldsymbol{c}^H(\tau)\boldsymbol{e}_i|^2} \right\} \tag{4.33}$$

where the $\boldsymbol{e}_i$ are the noise subspace eigenvectors and the maxima are found at $\tau = \tau_k$, $k = 1, \ldots, M$.

An alternative to this time domain approach is to take the DFT of (4.29) and operate in the frequency domain [120]. This reduces the problem to the identification of complex sinusoids. The Tufts-Kumerasan method [88], Root-MUSIC [35], and the total least squares version of ESPRIT (TLS-ESPRIT) [145] have all been used to estimate the time delays in this frequency domain approach.

Conventional code tracking. Code tracking circuits take over the synchronization process after initial acquisition has been achieved. They attempt to maintain a fine alignment between the incoming and local codes using feedback loops. Code tracking is accomplished using phase-lock techniques very similar to those used for generation of coherent carrier references. Two popular forms of code trackers are the delay lock loop (DLL) and the tau-dither loop (TDL) [163, 165]. The noncoherent DLL, shown in Fig. 4.4, uses correlation operations between the received signal and two different phases (early and late) of the local code generator. It operates by correlating the received signal with the early and late spreading codes $a(t - \hat{\tau} + \Delta \cdot T_c)$ and $a(t - \hat{\tau} - \Delta \cdot T_c)$, respectively, where $\hat{\tau}$ is an estimate of the delay between the local and incoming codes. The code phase error signal $e(t)$ is obtained by squaring and differencing the correlator outputs. Since the data modulated signal and carrier phase shift are unknown, in general, squaring the correlator outputs tends to remove their effects. The loop is closed by lowpass-filtering $e(t)$ with $f(t)$, and the output $u(t)$ is used to drive the voltage controlled clock (VCC) and correct the code phase error of the local code generator. The parameter Δ, $0 < \Delta < 1$, is called the *early-late discriminator offset.* The output of the VCC provides the channel time delay estimate, $\hat{\tau}$. The DLL is often referred to as the *full-time early-late tracking loop.*

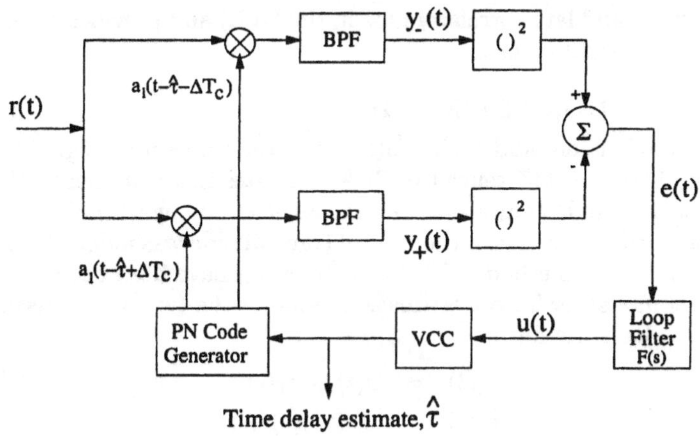

Figure 4.4. The noncoherent DLL for code tracking.

The DLL uses two independent channels (early and late) to perform the two correlations. An alternative is to use a single channel that is time shared and this forms the basis of the *tau-dither early-late tracking loop* [207]. It has the advantage of being more simple electronically than the DLL.

The tracking loops based on correlation techniques, such as those just discussed, are designed to operate in AWGN channels and thus do not operate well in multipath fading channels [151]. Consequently, modified code tracking loops have been developed that give improved performance in multipath channels. These techniques include a multipath estimating DLL (MEDLL) [185], a RAKE-DLL [152], and a 2-mode DLL [177].

Kalman filtering for code tracking. Besides making modifications to conventional code tracking loops, methods have been developed based on extended Kalman filtering (EKF) to provide robust code tracking in multipath channels [46, 76, 77]. The EKF [5] is used because estimation of propagation delays is highly nonlinear. The EKF-based methods are based on direct application of the EKF to the received signal given in (2.24). Unfortunately, the EKF may be too complex for some applications. Also, modeling errors as well as errors in the estimate of the number of resolvable paths can cause divergence in the algorithm [76].

A modification to the this approach was presented in [73] which assists the EKF by providing channel attenuation measurements from a RAKE receiver with maximal ratio combining. The EKF takes the outputs from early and late correlators (as in the DLL) and provides an estimate of the propagation delay.

1.3.2 TDOA ESTIMATION

An obvious method of forming TDoA estimates for a signal received at two BSs is to difference two ToA estimates made at those BSs. This approach would require that the BS receivers be synchronized to some common time reference if the time of transmission was unknown. However, more direct methods of TDoA estimation have been developed. The general model for TDoA estimation assumes the two signals received at the BSs have the form

$$r_1(t) = s(t) + n_1(t) \tag{4.34a}$$

$$r_2(t) = As(t-\tau) + n_2(t) \tag{4.34b}$$

where τ is the delay of the signal arriving at the second BS relative to the first BS. The goal is to estimate τ. In the following, a few of the many techniques that can be found in the literature are discussed.

Correlation methods. The conventional methods for computing these time estimates use cross-correlation techniques such as generalized cross-correlation (GCC) for TDoA estimation which finds the argument that maximizes the cross-correlation between signals received at a pair of BSs [85, 49, 50]. The cross-correlation function of the signals in (4.34a) and (4.34b) is given by

$$C_{1,2}(\tau) = \frac{1}{T}\int_0^T r_1(t)r_2(t+\tau)dt \ . \tag{4.35}$$

The TDoA estimate is the value $\hat{\tau}$ that maximizes $C_{1,2}(\tau)$. This approach requires that the analog signals $r_1(t)$ and $r_2(t)$ be digitized and transmitted to a common site for processing.

Improvement in the accuracy of the delay estimate, $\hat{\tau}$, can be obtained by filtering the two received signals prior to cross-correlation. To understand why filtering can improve the performance, consider the cross-power spectral density function, $U^r_{1,2}(f)$, which is related to the cross-correlation function in (4.35) by

$$C^r_{1,2}(\tau) = \int_{-\infty}^{\infty} U^r_{1,2} e^{j\pi f\tau} df \ . \tag{4.36}$$

If the signals $r_1(t)$ and $r_2(t)$ are filtered to produce $f_1(t)$ and $f_2(t)$, respectively, prior to cross-correlation, then the cross-power spectral density becomes

$$U^f_{1,2}(f) = H_1(f)H_2^*(f)U_{1,2}(f) \tag{4.37}$$

where $H_i(f)$ is the filter used for $r_i(t)$, $i = 1, 2$. Thus, we can express the generalized cross-correlation as

$$C^f_{1,2}(\tau) = \int_{-\infty}^{\infty} U^f_{1,2}(f) H_1(f) H_2^*(f) e^{j\pi f\tau} df \ . \tag{4.38}$$

Proper choice of the product $H_1(f)H_2^*(f)$ helps to minimize the effect of noise and interference. Details of various choices of the filters can be found in [21, 60, 85, 142].

Other methods such as Cyclic Cross-Correlation (CYC-COR), Spectral Coherence Alignment (SPECCOA), and Band-Limited Spectral Correlation Ratio (BL-SPECCOR) use cross-correlation methods that exploit the cyclostationarity of signals [48].

Other TDoA estimators. An extensive body of literature exists which addresses methods of high resolution TDoA estimation, far more than are able to be listed here. A few of these include estimators based on phase data [129], least squares [24, 105], least mean squares [111], and maximum likelihood [194]. A good source of further information is provided in [20] and [75].

2. RADIOLOCATION ALGORITHMS

Two approaches can be taken for solving for the location of a MS given measurements of signal strength, AoAs, ToAs or TDoAs. The straightforward approach is to use a geometric interpretation of the measurements and compute the intersections of the lines of position. This works if there are not measurement errors and the system is not overdetermined (more measurements available than there are unknowns). If errors are present, then the lines of position will not intersect at a single point, with the exception of the AoA method since two straight lines are used for location in two dimensions. Consequently, statistical approaches have been developed for these cases. In the following, geometric and statistical solutions to the location problem are developed. To simplify the mathematics, only location in two dimensions is considered. Extension to higher dimensions is straightforward and many of the references cited include results for three dimensions.

2.1 GEOMETRIC APPROACHES

Recall that the lines of position for ranging (signal strength and ToA) are circles, for TDoA they are hyperbolas, and for AoA they are straight lines. For ranging, we will only refer to the ToA method, although the results are the same for signal strength location systems.

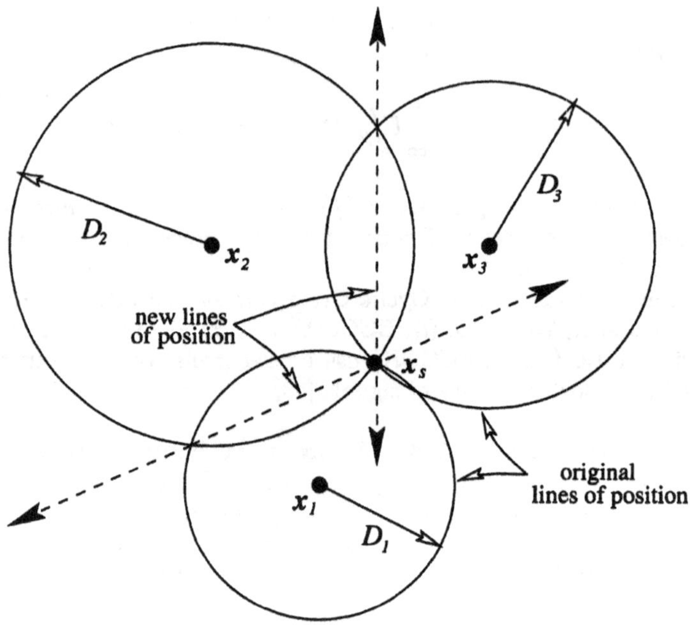

Figure 4.5. Intersection of circular lines of position from intersection of lines. Only two of the three lines are shown.

Given a ToA measured at BS i, the equation for a circular line of position is given by

$$t_i = D_i(\boldsymbol{x}_s)/c \tag{4.39}$$

where

$$\begin{aligned} D_i(\boldsymbol{x}_s) &= \|\boldsymbol{x}_i - \boldsymbol{x}_s\| \\ &= \sqrt{(x_i - x_s)^2 + (y_i - y_s)^2} \end{aligned} \tag{4.40}$$

is the measured range between the MS and BS, $\boldsymbol{x}_i$ is the position of ith BS, $\boldsymbol{x}_s$ is the position of the source (MS) and c is the speed of light. Since distance and time are directly related by the speed of light, the equations can be developed based on either time or distance. By using at least three BSs to resolve ambiguities, the MS's position is given by the intersection of circles.

The approach presented here for calculating the intersection of the circles is shown in Fig. 4.5. Each pair of circles can intersect in at most two points which can be used to define a straight line. The intersection of two of these lines provides the location of the MS. Thus, the lines define new lines of position. Squaring and differencing the equations in (4.39) for $i = 1, 2$ and solving for the y-coordinate of $\boldsymbol{x}_s$ gives

$$y_s = -\left(\frac{x_2 - x_1}{y_2 - y_1}\right) x_s + \frac{x_2^2 + y_2^2 - x_1^2 - y_1^2 + D_1^2 - D_2^2}{2(y_2 - y_1)} \tag{4.41}$$

where D_i is the measured range from the MS to the ith BS given in (4.40). Following the same procedure for $i = 2, 3$ yields

$$y_s = -\left(\frac{x_2 - x_3}{y_2 - y_3}\right) x_s + \frac{x_2^2 + y_2^2 - x_3^2 - y_3^2 + D_3^2 - D_2^2}{2(y_2 - y_3)} \ . \tag{4.42}$$

Equating these two linear equations and solving for x_s, then y_s, gives

$$x_s = \frac{(y_2 - y_1)\, C_3 - (y_2 - y_3)\, C_1}{2\left[(x_2 - x_3)(y_2 - y_1) - (x_2 - x_1)(y_2 - y_3)\right]} \tag{4.43}$$

$$y_s = \frac{(x_2 - x_1)\, C_3 - (x_2 - x_3)\, C_1}{2\left[(y_2 - y_3)(x_2 - x_1) - (y_2 - y_1)(x_2 - x_3)\right]} \tag{4.44}$$

where

$$C_1 = x_2^2 + y_2^2 - x_1^2 - y_1^2 - D_1^2 - D_2^2 \tag{4.45}$$

$$C_3 = x_2^2 + y_2^2 - x_3^2 - y_3^2 - D_3^2 - c^2 D_2^2 \ . \tag{4.46}$$

Thus, given the positions of the BSs and the range measurements, $D_i = c\tau_i$, an estimate of the MS's location can be obtained using (4.43) and (4.44). Also, note that this method can be used when there are measurement errors and the circles don't all intersect at a single point. The dilemma is which pair of lines to use.

In the TDoA approach, the measured time difference of arrival between BS i and j, with respect to BS j, is modeled by the hyperbolic equation

$$\rho_{i,j} = \frac{D_i(\boldsymbol{x}_s) - D_j(\boldsymbol{x}_s)}{c} \tag{4.47}$$

which has its foci at the two BSs. For location in two dimensions, three BSs are required which produces two TDoA pairs[1]. Consider TDoA

[1] In general, N_B BSs provide $N_B(N_B - 1)/2$ possible choices of BSs to form TDoAs. However, only $N_B - 1$ of the TDoAs are non-redundant, i.e., are independent. For instance, in two dimensions, we can have $\rho_{2,1}$, $\rho_{3,1}$ and $\rho_{3,2}$, but the last is a linear combination of the other two. Thus, there are only two independent TDoAs.

measurements made with respect to the first BS, $\rho_{2,1}$ and $\rho_{3,1}$, for two-dimensional location. Rearranging equation (4.47) and squaring yields

$$\begin{aligned} 2c\rho_{i,1}D_1(\boldsymbol{x}_s) &= D_i^2(\boldsymbol{x}_s) - D_1^2(\boldsymbol{x}_s) - c^2\rho_{i,1}^2 && (4.48)\\ &= \|\boldsymbol{x}_i\|^2 - \|\boldsymbol{x}_1\|^2 + 2x_s(x_1 - x_i)\\ &\quad +2y_s(y_1 - y_i) - c^2\rho_{i,1}^2 && (4.49) \end{aligned}$$

where $i = 2, 3$ and, without loss of generality, we have assumed $j = 1$. Solving for $D_1(\boldsymbol{x}_s)$ for $i = 2, 3$ and equating the results, it can be shown that [40]

$$y_s = mx_s + b \qquad (4.50)$$

where

$$\begin{aligned} m &= \frac{\rho_{3,1}(x_2 - x_1) - \rho_{2,1}(x_3 - x_1)}{\rho_{2,1}(y_3 - y_1) - \rho_{3,1}(y_2 - y_1)} && (4.51)\\ b &= \frac{\rho_{2,1}\|\boldsymbol{x}_3\|^2 - \rho_{3,1}\|\boldsymbol{x}_2\|^2 + \rho_{3,2}\|\boldsymbol{x}_1\|^2 + c^2\rho_{3,1}\rho_{2,1}\rho_{3,2}}{2\left[\rho_{2,1}(y_3 - y_1) - \rho_{3,1}(y_2 - y_1)\right]}. && (4.52) \end{aligned}$$

The term $\rho_{3,2}$ is not a measured TDoA but rather inserted for $\rho_{3,1} - \rho_{2,1}$ for notational convenience, as noted in the previous footnote. Substituting y_s back into (4.49) and solving for x_s yields the quadratic

$$Ax_s^2 + Bx_s + C = 0 \qquad (4.53)$$

where

$$\begin{aligned} A &= 4c^2\rho_{2,1}^2(1+m)^2 - u^2 && (4.54)\\ B &= -4c^2\rho_{2,1}^2\left[2x_1 - 2m(y_1 - b)\right] + 2u && (4.55)\\ C &= 4c^2\rho_{2,1}^2\left(\|\boldsymbol{x}_1\|^2 - 2y_1 b + b^2\right) - v^2 && (4.56) \end{aligned}$$

and

$$\begin{aligned} u &= 2\left[x_1 - x_2 + m(y_1 - y_2)\right] && (4.57)\\ v &= \|\boldsymbol{x}_2\|^2 - \|\boldsymbol{x}_1\|^2 + 2b(y_1 - y_2) - c^2\rho_{2,1}^2 && (4.58) \end{aligned}$$

which can then be used to compute y_s by equation (4.50). The solution to (4.53) yields two solutions for x_s leading to an ambiguity. The two solutions arise from the fact that there are two branches to the hyperbola. By using the squaring operations above, information regarding which branch the MS lies on has been lost. The ambiguity can be resolved by noting sign of one of the TDoAs, for instance $\rho_{2,1}$. If $\rho_{2,1} > 0$, then the MS lies closer to BS 1 than BS 2. By plugging both solutions into the $\rho_{2,1}$ defined in (4.47) and noting the sign, the correct solution can be found. Unfortunately, a consequence of this approach is that there is frequently two or more possible solutions and there is no way to distinguish them without more information [202]. For TDoA location in three dimensions, an extension of the approach used above was developed in [40] using four BSs. For the overdetermined case, in which there are more equations than unknowns, a method was developed in [66] for direct computation of a location estimate.

A further geometric interpretation of TDoA location was presented in [147]. Instead of the the range differences between two BSs providing a hyperbola on which the MS must lie, the new interpretation views the differences in range to three BSs as a straight line which is the major axis of a general conic, a focus of which is the location of the MS. The three BSs are on the conic whereas the foci of the conic are on the conic axis. Consequently, given four BSs, two general conics can be formed each of which have a foci at the position of the MS. The intersection of the conic axes for the two conics provide an estimate of the location of the MS in two dimensions. If only three BSs are available and the conic is an ellipse, the location of the MS can be determined by computing the foci and using the sign of a measured TDoA to choose the appropriate foci. If the conic is a hyperbola, an ambiguity occurs which must be resolved using additional information.

For AoA location, the true arrival angles at BS i can be modeled by

$$\phi_i = \tan^{-1}\left(\frac{y_i - y_s}{x_i - x_s}\right) \tag{4.59}$$

where it is assumed that each AoA is measured with respect to a common baseline, for instance the x-axis. Since a line can be defined by a point and an angle, the line of position for BS i is

$$y_s = x_s \tan\phi_i + (y_i - x_i \tan\phi_i) \ . \tag{4.60}$$

Equating the lines of position for the two BSs, $i = 1, 2$, and solving for x_s yields

$$x_s = \frac{y_2 - y_1 - x_2 \tan\phi_2 + x_1 \tan\phi_1}{\tan\phi_1 - \tan\phi_2} \tag{4.61}$$

which can be inserted back into (4.60) for $i = 1$ or $i = 2$ to form the estimate of y_s.

2.2 STATISTICAL APPROACHES

The geometric approaches can be used when there are errors in the ToA, TDoA or AoA estimates since the solution is obtained by direct substitution of the measured quantities. A better solution to the problem of location in the presence of measurement errors is to use a statistical approach. Such an approach is also useful for when the system is overdetermined. In this case, when the lines of position are straight lines, there may be multiple intersection points of those lines due to measurement errors. For lines of position that are not straight, again there will be multiple points of intersection, even when the system is not overdetermined. Consequently, statistical solutions provide a better means of estimating location.

2.2.1 PROBLEM FORMULATION

In general, the vector of noisy measurements, $\boldsymbol{r}_m$, from a set of N_B BSs can be modeled by

$$\boldsymbol{r}_m = \boldsymbol{C}(\boldsymbol{\theta}_s) + \boldsymbol{n}_m \tag{4.62}$$

where $\boldsymbol{n}_m$ is the measurement noise vector generally assumed to have zero mean and covariance matrix $\boldsymbol{\Sigma}_m$. $\boldsymbol{\theta}_s$ is the vector of parameters to be estimated and is generally equal to $\boldsymbol{x}_s$. The size of the vectors $\boldsymbol{r}_m$ and $\boldsymbol{n}_m$ are $N_B \times 1$ for ToA and AoA, and $(N_B - 1) \times 1$ for TDoA. The system measurement model $\boldsymbol{C}(\boldsymbol{\theta}_s)$ depends on the location method used:

$$\boldsymbol{C}(\boldsymbol{\theta}_s) = \begin{cases} \boldsymbol{D}(\boldsymbol{\theta}_s) & \text{for ToA} \\ \boldsymbol{\mathcal{R}}(\boldsymbol{\theta}_s) & \text{for TDoA} \\ \boldsymbol{\Phi}(\boldsymbol{\theta}_s) & \text{for AoA} \end{cases} \tag{4.63}$$

where

$$\boldsymbol{D}(\boldsymbol{\theta}_s) = [t_1, t_2, \ldots, t_N]^T \tag{4.64}$$

$$\boldsymbol{\mathcal{R}}(\boldsymbol{\theta}_s) = [\rho_{2,1}, \rho_{3,1}, \ldots \rho_{N,1}]^T \tag{4.65}$$

$$\boldsymbol{\Phi}(\boldsymbol{\theta}_s) = [\phi_1, \phi_2, \ldots, \phi_N]^T \ . \tag{4.66}$$

The terms t_i, $\rho_{i,1}$ and ϕ_i are the ToAs, TDoAs and AoAs defined in equations (4.39), (4.47) and (4.59), respectively. The TDoAs are referenced to the first BS. Although it is not explicitly shown, t_i, $\rho_{i,1}$ and ϕ_i are nonlinear functions of $\boldsymbol{\theta}_s$.

If the time of transmission t_s is needed to form the ToA estimates and it is not known, it can be incorporated into $\boldsymbol{\theta}_s$ as a parameter to be estimated along with x_s and y_s. The unknown parameter vector can then be modified to $\boldsymbol{\theta}_s = [\boldsymbol{x}_s^T, t_s]^T$ while the system measurement model becomes

$$\boldsymbol{C}(\boldsymbol{\theta}_s) = t_s\mathbf{1} + \boldsymbol{D}(\boldsymbol{x}_s)/c \tag{4.67}$$

where $\mathbf{1} = [1, 1, \ldots, 1]^T$ is the column vector of ones. In this case, t_i is indicative of a time-stamp of the signal's arrival and doesn't represent the measured range to the MS, which is actually $t_i - t_s$.

2.2.2 LOCATION SOLUTIONS

A common approach for determining an estimate from a noisy set of measurements is the method of LS estimation. The weighted least squares (WLS) solution is formed as the vector $\hat{\boldsymbol{\theta}}_s$ that minimizes a cost function

$$\mathcal{E}(\hat{\boldsymbol{\theta}}_s) = \left[\boldsymbol{r}_m - \boldsymbol{C}(\hat{\boldsymbol{\theta}}_s)\right]^T \boldsymbol{W} \left[\boldsymbol{r}_m - \boldsymbol{C}(\hat{\boldsymbol{\theta}}_s)\right] . \quad (4.68)$$

LS methods can achieve the ML estimate when the measurement noise vector is Gaussian with $\mathrm{E}[\boldsymbol{n}_m] = 0$ and equal variances [167], i.e., $\boldsymbol{\Sigma}_m = \sigma_n^2 \boldsymbol{I}$. If the variance are unequal, then WLS gives the ML estimate with with $\boldsymbol{W} = \boldsymbol{\Sigma}_m^{-1}$. Even when the noise cannot be assumed to be Gaussian, this is a reasonable approach for finding an estimator for $\boldsymbol{\theta}_s$. For the analysis that follows, we will assume that $\boldsymbol{W} = \boldsymbol{I}$.

As equation (4.63) indicates, $\boldsymbol{C}(\boldsymbol{\theta}_s)$ is a nonlinear function of the unknown parameter vector $\boldsymbol{x}_s$ so that the LS problem is a nonlinear one. One straightforward approach is to iteratively search for the minimum of the function using a gradient descent method [16]. With this approach, an initial guess is made of the MS location and successive estimates are updated according to

$$\hat{\boldsymbol{\theta}}_s^{(k+1)} = \hat{\boldsymbol{\theta}}_s^{(k)} - \boldsymbol{\mu}\, \nabla \mathcal{E}(\hat{\boldsymbol{\theta}}_s^{(k)}) \quad (4.69)$$

where the matrix $\boldsymbol{\mu} = \mathrm{diag}(\mu_x, \mu_y)$ is the step size, $\hat{\boldsymbol{\theta}}_s^{(k)}$ is the estimate at iteration k, and $\nabla = \partial / \partial \boldsymbol{\theta}$ denotes the gradient vector with respect to the vector $\boldsymbol{\theta}$. Because the system of equations is nonlinear, the error surface is multimodal. Consequently, the algorithm could converge to one of the local minima and not the global minimum.

In order to mold the problem into a linear LS problem, the nonlinear function $\boldsymbol{C}(\boldsymbol{\theta}_s)$ can be linearized using a Taylor series expansion about some reference point $\boldsymbol{\theta}_o$ so that [43, 180, 181]

$$\boldsymbol{C}(\boldsymbol{\theta}_s) \approx \boldsymbol{C}(\boldsymbol{\theta}_o) + \boldsymbol{H}\,(\boldsymbol{\theta}_s - \boldsymbol{\theta}_o) \quad (4.70)$$

where $\boldsymbol{H}$ is the Jacobian matrix of $\boldsymbol{C}(\boldsymbol{\theta}_s)$. Then, the LS solution can be formed as

$$\hat{\boldsymbol{\theta}}_s = \boldsymbol{\theta}_o + \left(\boldsymbol{H}^T \boldsymbol{H}\right)^{-1} \boldsymbol{H}^T \left[\boldsymbol{r}_m - \boldsymbol{C}(\boldsymbol{\theta}_o)\right] . \quad (4.71)$$

This approach can be performed iteratively, with each successive estimate being closer to the final estimate [43, 181]. The Taylor series approach introduces error when the linearized function $\boldsymbol{C}(\boldsymbol{\theta}_s)$ does not accurately approximate the nonlinear function. A key drawback to this LS approach is that a guess, $\boldsymbol{\theta}_o$, must be made of the MS's initial position.

For TDoA, several other algorithms have been developed that exploit the hyperbolic equations. One LS-based location algorithm, called the spherical interpolation (SI) method, was developed in [158]. By mapping the spatial origin to the first BS and through some simple algebraic manipulation, an error expression can be obtained (with $\boldsymbol{\theta}_s = \boldsymbol{x}_s$),

$$\boldsymbol{e} = \boldsymbol{\delta} - 2R_s\boldsymbol{r}_m - 2\boldsymbol{S}\boldsymbol{x}_s \tag{4.72}$$

where $R_s = \|\boldsymbol{x}_s\|$, $\boldsymbol{e} = [e_1, \ldots, e_{N_B}]^T$ are the errors, $\boldsymbol{r}_m = \boldsymbol{\mathcal{R}}$ are the TDoA measurements as defined in (4.63),

$$\boldsymbol{\delta} = \begin{bmatrix} R_s^2 - \rho_{2,1}^2 \\ \vdots \\ R_s^2 - \rho_{N_B,1}^2 \end{bmatrix} \quad \text{and} \quad \boldsymbol{S} = \begin{bmatrix} x_2 & y_2 \\ \vdots & \vdots \\ x_{N_B} & y_{N_B} \end{bmatrix}. \tag{4.73}$$

The error vector is the all zero vector if there are no measurement errors. LS minimization of this error is, of course, nonlinear so an alternative is chosen. Note that (4.72) is linear in $\boldsymbol{x}_s$ given R_s, and it is also linear in R_s given $\boldsymbol{x}_s$. Consequently, a two step LS solution is developed which first provides the linear LS solution of $\boldsymbol{x}_s$ given R_s, which is substituted back into (4.72) to form the linear LS solution of R_s given $\boldsymbol{x}_s$. The result of the method is the LS estimate

$$\hat{\boldsymbol{x}}_s = \frac{1}{2}\left[\boldsymbol{S}^T\boldsymbol{S}\right]^{-1}\boldsymbol{S}^T\left[\boldsymbol{\delta} - 2\hat{R}_s\boldsymbol{r}_m\right] \tag{4.74}$$

where $\hat{R}_s$ is the LS solution for R_s. A variation on this algorithm is called the spherical intersection (SX) method which inserts the linear LS estimate of $\boldsymbol{x}_s$ given R_s into $R_s = \boldsymbol{x}_s^T\boldsymbol{x}_s$ [157]. The result is a quadratic equation in R_s, the positive root of which is used to compute the location estimate.

Another LS approach that has been developed for TDoA to avoid linearization transforms the TDoA measurements into "pseudomeasurements" which are known functions of the actual TDoA measurements. The pseudomeasurements are given by [44]

$$\boldsymbol{\varphi} = \boldsymbol{\Delta}\boldsymbol{x}_s + D_1(\boldsymbol{x}_s)\boldsymbol{r}_m \tag{4.75}$$

where

$$\boldsymbol{\Delta} = \begin{bmatrix} (\boldsymbol{x}_2 - \boldsymbol{x}_1)^T \\ \vdots \\ (\boldsymbol{x}_N - \boldsymbol{x}_1)^T \end{bmatrix} \tag{4.76}$$

and

$$\varphi = \frac{1}{2}\begin{bmatrix} \|\boldsymbol{x}_2\|^2 - \|\boldsymbol{x}_1\|^2 - \rho_{2,1}^2 \\ \vdots \\ \|\boldsymbol{x}_N\|^2 - \|\boldsymbol{x}_1\|^2 - \rho_{N,1}^2 \end{bmatrix} . \tag{4.77}$$

We refer to the elements of the vector φ as the "pseudomeasurements" since they represent a transformation, through the function in equation (4.77), of the actual TDoAs, $\rho_{i,1}$. The term $D_1(\boldsymbol{x}_s)$ is nonlinear in the unknown vector $\boldsymbol{x}_s$ and can be removed by using a projection matrix that has $\boldsymbol{r}_m$ in its null space. A suggested projection is $\boldsymbol{P} = (\boldsymbol{I} - \boldsymbol{Z})[\text{diag}(\boldsymbol{r}_m)]^{-1}$ where $\boldsymbol{Z}$ is a circular shift matrix [44]. Projecting (4.75) with $\boldsymbol{P}$, the following linear equation results:

$$\boldsymbol{P}\varphi = \boldsymbol{P}\Delta\boldsymbol{x}_s \tag{4.78}$$

which leads to the linear LS solution for the location of the MS given by

$$\hat{\boldsymbol{x}}_s = \left(\Delta^T\boldsymbol{P}^T\boldsymbol{P}\Delta\right)^{-1}\Delta^T\boldsymbol{P}^T\boldsymbol{P}\varphi . \tag{4.79}$$

For location in n dimensions, this method requires $n + 2$ BSs. It was shown in [44] that this result is equivalent to that given in (4.74) for the SI method. A more general form for this LS solution was developed in [136]. Another two-stage LS minimization method was developed in [25] that is optimal for small TDoA estimation errors.

2.3 COMMENTS ON LOCATION ALGORITHMS

The algorithms presented in the previous section vary in their degree of accuracy and complexity. It has been found that the Taylor series expansion algorithms perform accurately when iterated which increases computational complexity. The SI, SX, "pseudomeasurement" and geometric algorithms generally do not perform as well as the Taylor series algorithm. The algorithm in [25] offers a more accurate closed form solution when the TDoA estimation errors are small. The general accuracy of TDoA source location has been treated in [27, 90, 106, 166]. For the algorithms described above, accuracy analyses are derived in their corresponding references.

To track the location of a moving MS, frequent iteration of the algorithms is necessary with new ToA, TDoA or AoA measurements required for each. More appropriate methods for tracking are based on recursive least squares (RLS) and Kalman filtering [180] which can often be applied to the measurement models employed in the estimation algorithms discussed above. Chapter 7 discusses the use of these methods in more detail.

Finally, it is important to note that each of the algorithms above assumes a direct, or LoS, path exists between the MS and each BS that is used for location. When this is not the case, significant error can be introduced into the location estimate. Consideration of the effect of NLoS propagation is addressed in Chapter 7.

3. MEASURES OF LOCATION ACCURACY

To evaluate the performance of a location method, several benchmarks have been proposed. A common measure of accuracy is the comparison of the mean-squared-error (MSE) or the *rms* error of the location estimate with the Cramér-Rao lower bound (CRLB) [186]. The concepts of circular error probability (CEP) [180] and geometric dilution of precision (GDOP) [108] have also been used as accuracy measures. These are briefly discussed below.

3.1 CRAMÉR-RAO LOWER BOUND

For location in D dimensions ($D = 2$ or 3), the MSE of the position estimate is given by

$$\text{MSE} = \text{E}\left[\left(\boldsymbol{\theta}_s - \hat{\boldsymbol{\theta}}_s\right)^T \left(\boldsymbol{\theta}_s - \hat{\boldsymbol{\theta}}_s\right)\right] \tag{4.80}$$

where $\text{E}[\cdot]$ denotes expectation. The root-mean-square (*rms*) error is simply $\sqrt{\text{MSE}}$. The calculated MSE is often compared to the theoretical minimum MSE given by the CRLB which sets a lower bound on the variance of any unbiased estimator. The CRLB is the inverse of the information matrix $\boldsymbol{J}$ defined as [186]

$$\boldsymbol{J} = \text{E}\left[\left(\frac{\partial p(\boldsymbol{r}_m|\boldsymbol{\theta})}{\partial \boldsymbol{\theta}}\right)\left(\frac{\partial p(\boldsymbol{r}|\boldsymbol{\theta})}{\partial \boldsymbol{\theta}}\right)^T\right]\Bigg|_{\boldsymbol{\theta}=\boldsymbol{\theta}_s} \tag{4.81}$$

where $\boldsymbol{r}_m$ is the vector of TDoA, ToA or AoA measurements and $p(\boldsymbol{r}_m|\boldsymbol{\theta})$ is the probability density function of $\boldsymbol{r}_m$ conditioned on the parameter vector $\boldsymbol{\theta}$. Assuming Gaussian measurement noise, $p(\boldsymbol{r}_m|\boldsymbol{\theta})$ is Gaussian with mean $\boldsymbol{r}_0$ and covariance matrix $\boldsymbol{Q}$, and the CRLB reduces to

$$\text{CRLB} = \boldsymbol{J}^{-1} = c^2 \left(\frac{\partial \boldsymbol{r}_0^T}{\partial \boldsymbol{\theta}} \boldsymbol{Q}^{-1} \frac{\partial \boldsymbol{r}_0}{\partial \boldsymbol{\theta}^T}\right)^{-1}\Bigg|_{\boldsymbol{\theta}=\boldsymbol{\theta}_s} . \tag{4.82}$$

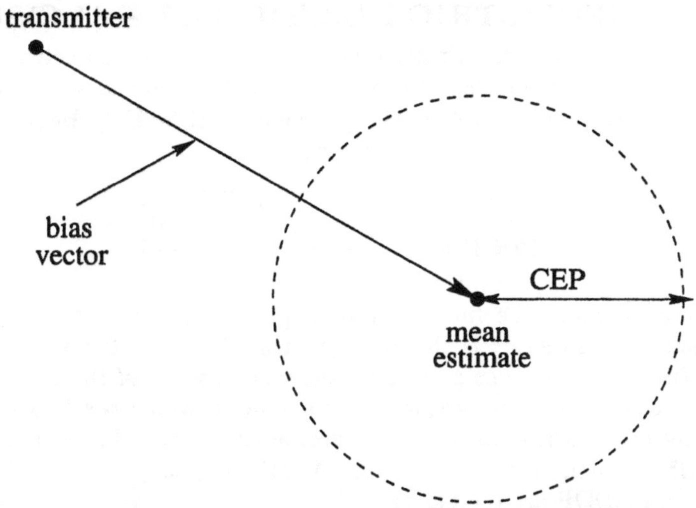

Figure 4.6. The geometry of the CEP measure of accuracy.

3.2 CIRCULAR ERROR PROBABILITY

A simple measure of accuracy is the CEP which is defined as the radius of the circle that has its center at the mean and contains half the realizations of a random vector. The CEP is a measure of the uncertainty in the location estimator $\hat{\boldsymbol{\theta}}_s$ relative to its mean $\mathrm{E}\left[\hat{\boldsymbol{\theta}}_s\right]$. If the location estimator is unbiased, the CEP is a measure of the estimator uncertainty relative to the true transmitter position. If the magnitude of the bias vector is bounded by B, then with a probability of one-half, a particular estimate is within a distance of B+CEP from the true position. The concept of the CEP measure is illustrated in Fig. 4.6.

Because the CEP is difficult to derive an exact expression for, an approximation that is accurate to within 10% is often used. The approximation for CEP is given as [180]

$$\mathrm{CEP} \approx 0.75\sqrt{\mathrm{E}\left[\left(\hat{\boldsymbol{\theta}}_s - \hat{\boldsymbol{\mu}}\right)^T \left(\hat{\boldsymbol{\theta}}_s - \hat{\boldsymbol{\mu}}\right)\right]} \tag{4.83}$$

$$= 0.75\sqrt{\sum_{i=1}^{D} \sigma^2_{\hat{\theta}_{s,i}}} \tag{4.84}$$

where $\hat{\boldsymbol{\mu}} = \mathrm{E}\left[\hat{\boldsymbol{\theta}}_s\right]$ is the mean location estimate and $\sigma^2_{\hat{\theta}_{s,i}}$ is the variance of the ith estimated coordinate.

3.3 GEOMETRIC DILUTION OF PRECISION

The GDOP provides a measure of the effect of the geometric configuration of the BSs on the location estimate. It is defined as the ratio of the *rms* position error to the *rms* ranging error [108, 180]. Hence, for an unbiased estimator, the GDOP is given by

$$\mathrm{GDOP} = \frac{\sqrt{\mathrm{E}\left[(\hat{\boldsymbol{x}}_s - \hat{\boldsymbol{\mu}})^T (\hat{\boldsymbol{x}}_s - \hat{\boldsymbol{\mu}})\right]}}{\sigma_r} \tag{4.85}$$

where σ_r denotes the fundamental ranging error for ToA and TDoA systems. For AoA, σ_r^2 is the average variance of the distance between each BS and a reference point near the true position of the MS.

The GDOP is an indicator of the extent to which the fundamental ranging error is magnified by the geometric relation between the MS and BSs. Furthermore, comparing (4.84) and (4.85), we find that the CEP and GDOP are related by

$$\mathrm{CEP} \approx (0.75\sigma_r)\mathrm{GDOP} \ . \tag{4.86}$$

The GDOP serves as a useful criterion for selecting the set of BSs from a large set to produce the minimum location error. In addition, it may aid cell site planning for cellular networks which plan to provide location services to their users.

4. SUMMARY

This chapter has provided a brief introduction to radiolocation techniques for wireless systems. Several methods were discussed which provide the location parameters for the signal strength, AoA and time-based location methods. A multitude of AoA methods exist in the literature. Conventional methods for ToA and TDoA estimation rely on correlation techniques. Both geometric and statistical location algorithms were developed and discussed. The statistical methods are more appropriate for cases when there are considerable measurement errors of the location parameters or when the system is overdetermined. The method of LS appears to be the most popular form of TDoA estimator. Finally, a few measures of the location accuracy of a system were presented including the CRLB, CEP and GDOP.

Chapter 5

PERFORMANCE OF LOCATION IN CDMA SYSTEMS

The regulations set forth by the FCC for the location accuracy of E-911 calls are applicable to both existing and future MS handsets. Due to the large number of existing handsets in use, it is desirable for a wireless location system to function without requiring modifications to those handsets. Consequently, radiolocation using reverse link signaling has been the primary consideration for implementing a location system in present cellular systems. Most often, a reverse link location system is separate from the wireless network in that it is "overlayed" on top of the network. This allows the necessary complexity to be contained in the location system without overly burdening the cellular network. The overlay approach is necessary for current AMPS and IS-54 systems although current CDMA and GSM systems provide system messages that can be used for location[1].

A distinguishing characteristic of wireless location in a cellular system is the presence of a harsh propagation environment. Signal variability, which can be tens of decibels over distances as small as a wavelength, and the absence of a direct propagation path between the MS and BS both compound the problem of radiolocation. On the other hand, the more information that is gathered regarding the location of the MS, the more accurate the location estimate is expected to be. Hence, the use of more than the minimum number of BSs for location may improve

[1]The current revision of IS-95, the standard for CDMA systems in North America, provides the ability to locate MSs using forward link signaling by using information from pilot measurement messages to form TDoA pairs. Unfortunately, the accuracy of those TDoA measurements is limited to a chip period (equivalent to approximately 240 m) which drastically limits the accuracy of a location system based on those TDoAs. A similar capability is found in GSM systems. See Chapter 8 for more details.

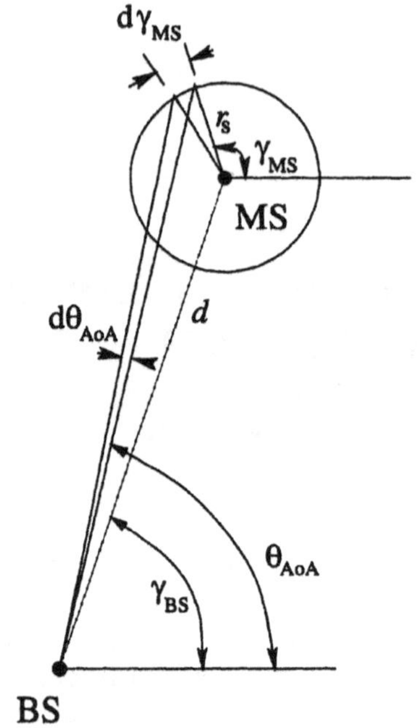

Figure 5.1. MS-BS geometry assuming a ring of scatterers for macrocells.

location accuracy. The influence of these factors on the performance of radiolocation systems that use reverse link signaling to estimate the ToAs and AoAs is quantified in this chapter where only the location of a single user is considered. The effects of other users are considered in a later chapter. The estimation of the ToA parameters are derived from the conventional method of the NC-DLL.

1. RADIOLOCATION SYSTEM

1.1 ANGLE OF ARRIVAL

For AoA systems, we consider the error due to multipath propagation but do not consider angle estimation errors due to algorithm assumptions or hardware issues. Multipath propagation, in the form of scattering near and around the MS and BS, will affect the measured AoA as discussed in Chapter 3. Jakes [80] and Gans [47] have modeled the macrocellular propagation environment as a ring of scatterers about the MS, with the BS well outside the ring. Fig. 5.1 illustrates the geometry, where the primary scatterers are assumed to be on a ring of radius r_s about the MS. The distance between the BS and MS, d, is assumed to be much greater than r_s. We assume that the MS uses an omnidirectional antenna, so that the probability distribution function (*pdf*) of the transmitted rays is given by

$$p(\gamma_{\text{MS}}) = \frac{1}{2\pi}, \qquad 0 \leq \gamma_{\text{MS}} < 2\pi \ . \tag{5.1}$$

The distribution of the AoA at the BS, θ_{AoA}, is given by

$$\frac{\partial \gamma_{\text{MS}}}{\partial \theta_{\text{AoA}}} p(\theta_{\text{AoA}}) = 2p(\gamma_{\text{MS}}) \ . \tag{5.2}$$

From the geometry of Fig. 5.1, we find that [47]

$$\partial \gamma_{\text{MS}} \cong \left[(\frac{r_s}{d})^2 - (\gamma_{\text{BS}} - \theta_{\text{AoA}})^2 \right]^{-\frac{1}{2}} \partial \theta_{\text{AoA}} \ . \tag{5.3}$$

Therefore, $p(\theta_{\text{AoA}})$ is given by

$$p(\theta_{\text{AoA}}) = \begin{cases} K_{\text{AoA}} \left[(\frac{r_s}{d})^2 - (\gamma_{\text{BS}} - \theta_{\text{AoA}})^2 \right]^{-\frac{1}{2}} & , \gamma_{\text{BS}} - \theta_M \leq \\ & \quad \theta_{\text{AoA}} \leq \gamma_{\text{BS}} + \theta_M \\ 0 & , \text{otherwise} \end{cases} \tag{5.4}$$

where

$$\begin{aligned} \theta_M &= \arctan(r_s/d) \\ K_{\text{AoA}} &= \frac{1}{2 \arcsin(\frac{d}{r_s} \theta_M)} \ . \end{aligned}$$

θ_M is the maximum angle deviation at the BS for a given distance between MS and BS and scattering ring radius. Note that for $d \gg r_s$, a small angle approximation can be invoked, with the result that $\theta_M \approx r_s/d$ and $K_{\text{AoA}} \approx 1/\pi$.

The model $p(\theta_{AoA})$ provides the AoA distribution for signals arriving at a BS. The model assumed goes one step further by assuming that the *measured* AoA at a BS also has the distribution $p(\theta_{\text{AoA}})$. Since the measured angles are not equal to the true angles to the MS, the lines of position from the BSs will not intersect at the same point. This problem is resolved by deriving the location estimate from the centroid of the set of points defined by the intersecting lines of position. With three BSs, for example, the lines of position intersect at three points, $\boldsymbol{x}_1$, $\boldsymbol{x}_2$ and $\boldsymbol{x}_3$, where $\boldsymbol{x}_i = [x_i, y_i]^T$ is the vector containing the coordinates of the points and $\boldsymbol{x}^T$ denotes the transpose of $\boldsymbol{x}$. The location estimate $\hat{\boldsymbol{x}}_s$ is obtained by averaging the coordinates of the points of intersection, i.e., $\hat{\boldsymbol{x}}_s = \sum_{i=i}^{3} \boldsymbol{x}_i/3$.

It is possible to use other scattering models, such as those in Chapter 2, and use the AoA distribution derived in the associated references. However, the model used here was chosen for its mathematical simplicity. The primary difference will be in how the *pdf* peaks around the true AoA.

1.2 TIME OF ARRIVAL

Because of the focus on CDMA cellular radio systems in this book, methods for determining the times of arrival from the spread spectrum signal are of interest. As discussed in Chapter 4, the two methods for determining time delays in spread spectrum communications systems are code acquisition with a sliding correlator or matched filter, and code tracking with a DLL or TDL [207]. Since the DLL finely tracks the time delay, it is better suited for a location system. The DLL is an essential part of time estimation used for GPS, and provides reasonable accuracy over the satellite-earth propagation channel. Here, the DLL-based location system is investigated for its performance in cellular propagation environments.

The operation of the noncoherent DLL, shown in Fig. 4.4, for code tracking in spread spectrum systems is discussed in Chapter 4. It produces an estimate of the delay, $\hat{\tau}$, at the output of the VCC. With synchronized BSs, $\hat{\tau}$ will provide a useful parameter for the location algorithm of the next section.

2. TIME-BASED LOCATION ALGORITHMS

As discussed in Chapter 4, two approaches are generally used to calculate the location of a MS from ToA or TDoA estimates. One approach uses a geometric interpretation to calculate the intersection of circles or hyperbolas, depending on whether ToA or TDoA is used. This approach becomes difficult if the hyperbolas or circles do not intersect at a point due to time measurement errors. A second approach calculates the position using a LS solution [43, 62, 181], which is a more statistically justifiable approach.

The algorithm to be developed will be based on the ToA method with ToA estimates provided by the DLL. The time of transmission will be included in the estimation process, as in Chapter 4. The algorithm assumes that the MS, located at $\boldsymbol{x}_s$, transmits its sequence at time t_s. The N_B BS receivers located at coordinates $\boldsymbol{x}_1, \boldsymbol{x}_2, \ldots, \boldsymbol{x}_{N_B}$ receive the sequence at times $t_1, t_2, \ldots, t_{N_B}$. The relative geometry of the MS and BSs is shown in Fig. 5.2. From Chapter 4, let $\boldsymbol{\theta}_s = [\boldsymbol{x}_s^T, t_s]^T$ be the vector of unknowns to be estimated and consider the following function which

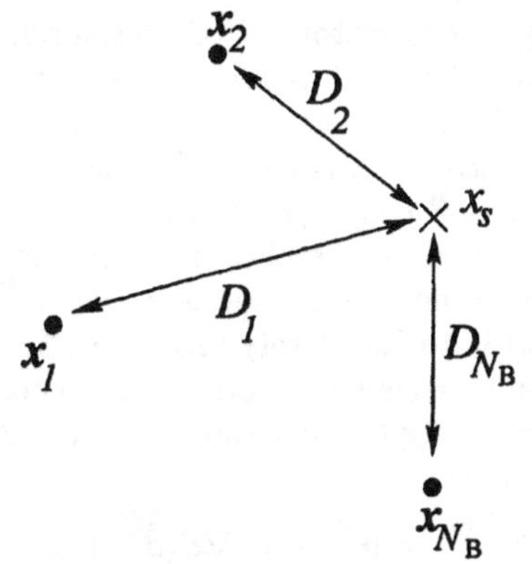

Figure 5.2. Geometry of the BSs, located at $\boldsymbol{x}_i$, $i = 1, \ldots, N_B$, in relation to the MS located at $\boldsymbol{x}_s$.

relates the unknown location and time of transmission to the position of each BS:

$$\boldsymbol{T}(\boldsymbol{\theta}_s) = t_s \mathbf{1} + \boldsymbol{D}(\boldsymbol{x}_s)/c \tag{5.5}$$

where $\boldsymbol{D}(\boldsymbol{x}_s) = [D_1(\boldsymbol{x}_s), \ldots, D_{N_B}(\boldsymbol{x}_s)]^T$ is a vector containing the distance between the ith BS and the MS defined in (4.40) and $\boldsymbol{T}(\boldsymbol{\theta}_s) = [T_1(\boldsymbol{\theta}_s), \ldots, T_{N_B}(\boldsymbol{\theta}_s)]^T$ for $i = 1, \ldots, N_B$. The operator $\|(\cdot)\|$ denotes the norm of the vector $(\cdot)$. Also, c is the speed of light and $\mathbf{1}$ represents the column vector of ones.

The functions, $T_i(\boldsymbol{\theta}_s)$, are formed for each BS receiver, $i = 1, \ldots, N_B$, and the differences $t_i - T_i(\boldsymbol{x}_s)$ could be made zero with the proper choice of $\boldsymbol{x}_s$ and t_s. However, the measured values of the arrival times t_i are generally in error due to multipath and other impairments, and NLoS propagation introduces errors into the range estimates that are derived from the arrival times.

2.0.1 UNCONSTRAINED NL-LS ALGORITHM

Since the geometric approach to finding the location of the MS fails when there are errors in the measured ToAs, statistical methods are used which, in general, are based on the method of least squares (LS). The LS methods produce the estimate $\hat{\boldsymbol{\theta}}_s = [\hat{\boldsymbol{x}}_s^T, \hat{t}_s]^T$ that minimizes the quadratic cost function

$$\begin{aligned} \mathcal{E}(\boldsymbol{\theta}_s) &= [\boldsymbol{t} - \boldsymbol{T}(\boldsymbol{\theta}_s)]^T \boldsymbol{W} [\boldsymbol{t} - \boldsymbol{T}(\boldsymbol{\theta}_s)] \\ &= \boldsymbol{e}^T(\boldsymbol{\theta}_s) \boldsymbol{W} \boldsymbol{e}(\boldsymbol{\theta}_s) \end{aligned} \tag{5.6}$$

where $\boldsymbol{t} = [t_1, \ldots, t_{N_B}]^T$ is the vector of ToA measurements from the N_B BSs. The matrix $\boldsymbol{W}$ is a diagonal matrix of weights α_i^2, $i = 1, \ldots, N_B$, that can be chosen to influence the contribution of the measurement from the ith BS on the location estimate. This will prove to be a useful method for mitigating NLoS propagation errors.

The method developed here searches directly for the minimum of $\mathcal{E}(\boldsymbol{\theta}_s)$ resulting in a nonlinear least squares (NL-LS) problem. One straightforward method is to iteratively search for the minimum of the function using a gradient descent method, where successive location estimates are updated according to the recursion in (4.69), which is reproduced here for convenience:

$$\hat{\boldsymbol{\theta}}_s^{(k+1)} = \hat{\boldsymbol{\theta}}_s^{(k)} - \boldsymbol{\mu}\, \nabla\mathcal{E}(\hat{\boldsymbol{\theta}}_s^{(k)}) \tag{5.7}$$

where the matrix $\boldsymbol{\mu}$ is the step size, $\hat{\boldsymbol{\theta}}_s^{(k)}$ is the estimate at time k, and $\nabla = \partial/\partial\boldsymbol{\theta}_s$ denotes the gradient vector with respect to the vector $\boldsymbol{\theta}_s$. Therefore,

$$\nabla\mathcal{E}(\hat{\boldsymbol{\theta}}_s^{(k)}) \equiv \nabla\mathcal{E}(\boldsymbol{\theta}_s)|_{\hat{\boldsymbol{\theta}}_s^{(k)}} = \begin{bmatrix} \left.\frac{\partial\mathcal{E}}{\partial x_s}\right|_{\hat{x}_s^{(k)}} \\ \left.\frac{\partial\mathcal{E}}{\partial y_s}\right|_{\hat{y}_s^{(k)}} \\ \left.\frac{\partial\mathcal{E}}{\partial t_s}\right|_{\hat{t}_s^{(k)}} \end{bmatrix}. \tag{5.8}$$

Since the difference $t_i - t_s$ is small (microseconds) compared to x_s and y_s (meters), the scalar step size $\boldsymbol{\mu}$ should be small enough to allow t_s to converge to a solution. Consequently, $\boldsymbol{\mu}$ is chosen to be the diagonal matrix

$$\mu = \begin{bmatrix} \mu_x & 0 & 0 \\ 0 & \mu_y & 0 \\ 0 & 0 & \mu_t \end{bmatrix} \tag{5.9}$$

where $\mu_x, \mu_y \gg \mu_t$. The recursion in (5.7) continues until $\|\nabla\mathcal{E}(\hat{\boldsymbol{\theta}}_s^{(k)})\|$ is smaller than some prescribed tolerance.

One drawback of the steepest descent method is its slow convergence. Since $\boldsymbol{T}(\boldsymbol{\theta}_s)$ is a nonlinear function of $\boldsymbol{x}_s$, linearization using a Taylor series expansion has been used to facilitate a unique global minimum of $\mathcal{E}(\boldsymbol{\theta}_s)$ [181, 43, 180]. The minimum of (5.6) is formed by linearizing $\boldsymbol{e}(\boldsymbol{\theta}_s)$ with a Taylor series expansion about $\hat{\boldsymbol{\theta}}_s^{(k)}$ and keeping only the first order terms, i.e.,

$$\boldsymbol{e}(\boldsymbol{\theta}_s) \approx \boldsymbol{e}(\hat{\boldsymbol{\theta}}_s^{(k)}) + \boldsymbol{\delta}^T\nabla\boldsymbol{e}(\hat{\boldsymbol{\theta}}_s^{(k)}) \tag{5.10}$$

where $\boldsymbol{\delta} = [\delta_x, \delta_y, \delta_t]^T = \boldsymbol{\theta}_s - \hat{\boldsymbol{\theta}}_s^{(k)}$. Substituting (5.10) into (5.6) and solving

$$\nabla\mathcal{E}(\boldsymbol{\theta}_s) = \boldsymbol{0} \tag{5.11}$$

for $\boldsymbol{\delta}$, the vector $\boldsymbol{\theta}_s^{(k)}$ is updated by

$$\hat{\boldsymbol{\theta}}_s^{(k+1)} = \hat{\boldsymbol{\theta}}_s^{(k)} + \boldsymbol{\delta} \ . \tag{5.12}$$

The gradient in (5.11) is with respect to the vector $\boldsymbol{\delta}$. The new estimate is substituted back into (5.10) and the process is reiterated until $|\delta_x| + |\delta_y| + c|\delta_t|$ is less than some prescribed tolerance.

When the MS is either close to the BSs or near the perimeter of the area defined by the polygon with the BSs as its vertices, then the linear approximation approach has convergence problems [62, 181]. For microcells, the MS is always within a short distance of the serving BS, so this method is not appropriate. The convergence problem arises from the approximation of $\boldsymbol{e}(\boldsymbol{\theta}_s)$ with the linear terms of the Taylor series expansion. Other objective functions $\mathcal{E}(\boldsymbol{\theta}_s)$ can be formed replacing, for example, the quadratic cost function with an absolute value function. However, these methods usually do not perform as well as minimizing the sum of squares [62].

2.0.2 CONSTRAINED NL-LS ALGORITHM

It may be possible to improve the time-based location algorithm due to the fact that the range error is always positive [115]. This is because the ToA estimates are always greater than the true ToA values due to multipath propagation and other impairments. Also, the range estimates derived from the ToA estimates are greater than the true ranges due to NLoS propagation. Therefore, the true location of the MS must lie inside the circles of radius $r_i = c(t_i - t_s)$, $i = 1, \ldots, N_B$, about the N_B BSs, since the MS cannot lie farther from a BS than its corresponding range estimate (Fig. 5.3). Mathematically, this implies

$$r_i = c(t_i - t_s) \geq \|\boldsymbol{x}_i - \boldsymbol{x}_s\| \ . \tag{5.13}$$

Since the unconstrained NL-LS algorithm does not take this restriction into account, a constrained NL-LS approach can be used to force the estimate at each iteration to satisfy equation (5.13). However, the LS solution is complicated by the nonlinear functionals $\boldsymbol{e}(\boldsymbol{\theta}_s)$ as well as the nonlinear inequality constraints of (5.13). Note that (5.13) implies that

$$\|\boldsymbol{x}_i - \boldsymbol{x}_s\|/c - (t_i - t_s) \leq 0 \ . \tag{5.14}$$

We recognize from (5.6) that the left side of the inequality in (5.14) is simply the ith component of $\boldsymbol{g}(\boldsymbol{\theta}_s) = -\boldsymbol{e}(\boldsymbol{\theta}_s)$. Hence, the restrictions $\boldsymbol{e}(\boldsymbol{\theta}_s) \geq 0$ are formed, where the area within the constraint boundaries is known as the *feasible region.*

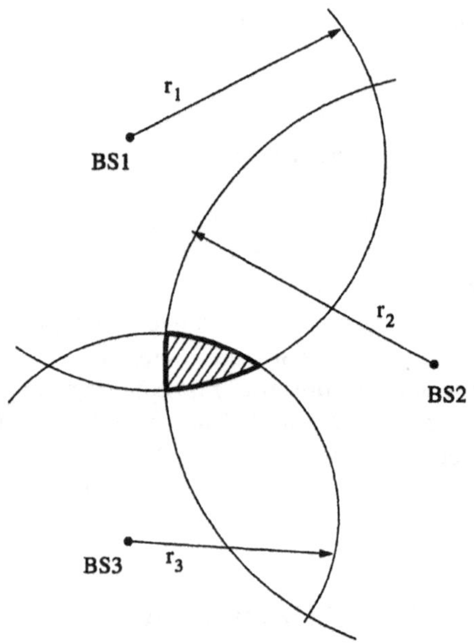

Figure 5.3. The location of the MS is constrained to the intersection area (shaded region) of circles of radius $c(t_i - t_s)$ centered at each BS.

There are many approaches to forming numerical solutions for NL-LS problems with nonlinear inequality constraints of the form $\boldsymbol{g}(\boldsymbol{\theta}_s) \leq 0$ [11]. One simple, yet effective, method uses penalty functions to modify the objective function $\mathcal{E}(\boldsymbol{\theta}_s)$ and form a solution using an unconstrained approach. The penalty functions provide a large penalty to the objective function when one or more of the constraints are violated. The objective function in (5.6) is modified to include the penalty functions $\boldsymbol{g}(\boldsymbol{\theta}_s)$ as follows [11]:

$$\mathcal{E}(\boldsymbol{\theta}_s) = \boldsymbol{e}^T(\boldsymbol{\theta}_s)\boldsymbol{W}\boldsymbol{e}(\boldsymbol{\theta}_s) - P \cdot \text{trace}\left\{\text{diag}\left[\boldsymbol{e}(\boldsymbol{\theta}_s)\right]^{-1}\right\} \qquad (5.15)$$

where P is positive for minimization and diag$[\boldsymbol{m}]^{-1}$ denotes the inverse of the diagonal matrix with elements from the vector $\boldsymbol{m}$. As any constraint is approached during the search, the penalty term forces $\mathcal{E}(\boldsymbol{\theta}_s)$ toward infinity, thus forming a natural optimum within the feasible region. This approach requires that the initial guess be placed within the feasible region. A method for doing this is described in [11].

The search procedure can be viewed as the optimization of a sequence of surfaces which tend toward the true value of the objective function.

Initially, an unconstrained search method is used to provide an artificial optimum $\theta_s^{(1)}$ with a large value of $P = P_1$. The next stage is initialized with the previous estimate $\theta_s^{(1)}$ and uses a smaller $P = P_2$ to provide a better approximation to the true optimum. In this way, the solution approaches the constraints more closely, if the optimum happens to lie close to one of the constraints. The penalty constraints become smaller at each stage, forming a monotonic-decreasing sequence $P_1 > P_2 > \ldots$, and the sequence of artificial optima $\theta_s^{(1)}, \theta_s^{(2)}, \ldots$, tends toward the true optimum. The search continues until several iterations fail to produce a change in the objective function. This formulation essentially replaces a constrained optimization by a sequence of unconstrained optimizations.

The unconstrained and constrained NL-LS algorithms just described will be used in the following sections to calculate a MS's location from the measured arrival times. If the MS is located at $\boldsymbol{x}_s$, then the location error is

$$e = \|\boldsymbol{x}_s - \hat{\boldsymbol{x}}_s\| \tag{5.16}$$

where $\hat{\boldsymbol{x}}_s$ is the estimate of the MS's location.

3. PROPAGATION MODELS

The three stage model for the radio propagation environment (including multipath fading, shadowing, and path loss) described in Chapter 2 is used with parameters described below.

The cell deployment for the macrocell simulations is shown in Fig. 5.4 where a BS and its 6 nearest neighbor BSs are separated by 6000 m. The wideband channel model uses 6 taps ($M = 6$) where the first tap delay, τ_0, is determined from the MS-BS geometry of Fig. 5.1 by calculating the distance traveled by a signal transmitted from the MS in a random direction according to $p(\gamma_{\text{MS}})$ and reflected from the ring of scatterers to the BS. The remaining delays are chosen according to the 6-tap reduced typical urban delay profile defined in COST 207 [30] (see Table 5.1). The model deviates slightly from the COST 207 model by assuming a classical Doppler spectrum for all taps, i.e., in the simulations the taps gains are all generated by using Jakes'multipath fading,Jakes' fading simulator method [80] with 8 oscillators. The shadowing model in (2.3) assumes a shadow decorrelation of 0.1 at a distance of 30 m. Finally, Hata's model for a medium or small city is used with a carrier frequency of $f_c = 850$ MHz, BS antenna heights of 100 m, and a MS antenna height of 2.5 m.

For microcells, the wideband channel model is also used with $M = 4$. Due to the site-specific nature of the microcellular propagation environment, techniques such as ray tracing have been developed. In this study,

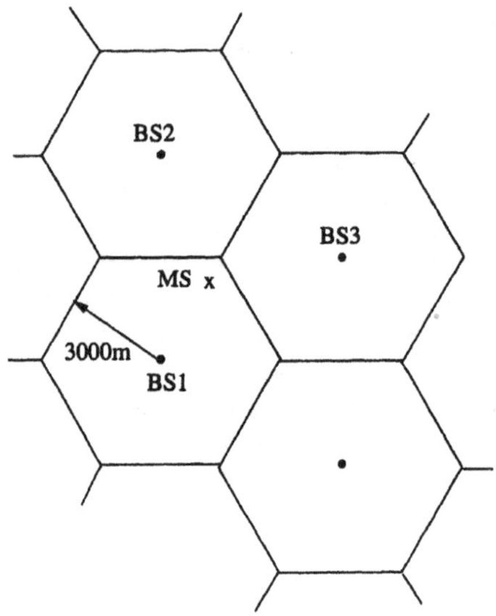

Figure 5.4. Macrocell deployment using typical hexagonal cells of radius 3000 m.

Table 5.1. COST 207 6-tap reduced typical urban power delay profile.

COST 207 Model	
Delay-τ_0 (μs)	*Fractional Power*
0.0	0.189
0.2	0.379
0.5	0.239
1.6	0.095
2.3	0.061
5.0	0.037

ray tracing concepts are used to calculate the propagation delays for the wideband channel model. A Manhattan street microcell BS deployment is assumed as shown in Fig. 5.5. When the MS is LoS with a BS, a 4-path model is used, consisting of a direct path, a road reflected path, and two wall reflected paths. The taps of the wideband channel model are generated using Jakes' method, appropriately modified for Rician fading. When the MS is NLoS with a BS, i.e., around the corner, a different approach is taken to determine the propagation delays. Since the literature provides no results that describe the power delay profile for a MS that is around a corner from a BS, a simplistic model is chosen.

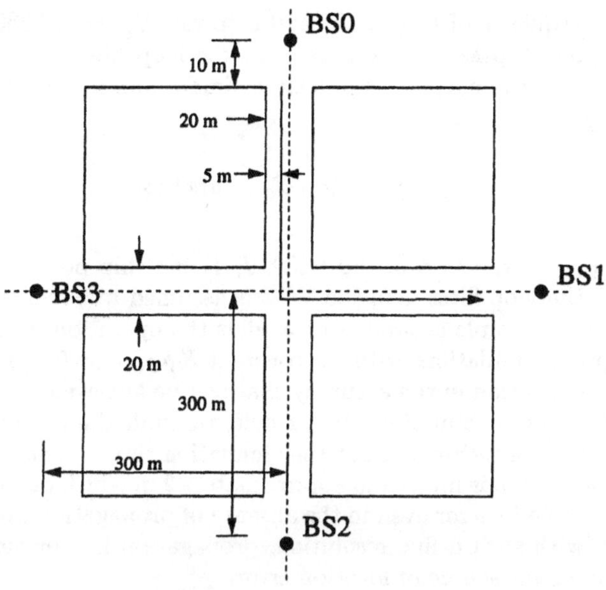

Figure 5.5. Manhattan street microcell deployment.

A 4-path NLoS propagation model is used that includes two paths that arrive from diffractions at the building corners in the street intersection and two remaining paths whose delays are generated by adding random delays to the first two paths. All paths are assumed to be Rayleigh faded. The model chosen here is inconsequential, because the extra time delay for NLoS BSs introduces a large amount of error into the location algorithm. Hence, accurate modeling of multipath propagation on NLoS streets is not necessary; only a means of introducing the excess propagation delay around the street corner is needed for our purposes.

The two-slope path loss model in (2.12) is used for the microcells with the breakpoint at $g = 150$ m and slopes before and after the break point of 2 ($m_a = 2$) and 4 ($m_b = 2$), respectively. The distance to the corner from each BS, as shown in Fig. 5.5, is approximately $d_c = 290$ m. The shadowing parameters are assumed to be the same as in the macrocell deployment.

4. SIMULATIONS

The location techniques described earlier in the chapter were simulated in the macrocellular and microcellular environments described in Chapter 2 to determine their performance. The spreading code used was an m-sequence of length 127 and chip rate $T_c^{-1} = 1.2288$ Mcps. In the DLL, an all-pass filter was used for the loop filter (i.e., $F(s) = 1$). For the VCC, the output time delay estimate and input waveform are related by

$$\hat{\tau}(t) = K_{VCC}T_c \int_0^t u(x)\, dx \tag{5.17}$$

where K_{VCC} is the gain of the VCC, T_c is the chip period, $u(t)$ is the output of the loop filter, and the VCC is assumed to begin operating at time $t = 0$. A simple accumulator models the operation of the VCC in the computer simulations with the constant $K_{VCC}T_c = 0.003$. Note that there is a limitation in the accuracy that can be achieved when simulating the DLL on a computer. As a result, we limit the resolution of the DLL to 1/120 of a chip to limit the simulation time. Consequently, the ranging resolution is limited to approximately 2 m which causes all range estimates to be in error even in the absence of propagation impairments. However, with such a fine resolution, propagation impairments will be the predominant source of location error.

4.1 RANGE ESTIMATION

Ranging measures the 1-D distance between a MS and a BS. Only the time-based method is employed for ranging since AoA ranging does not make sense. For macrocells, our ranging results assume that the first path to arrive from the COST 207 model is a LoS path. Consequently, the ranging results for macrocells are very optimistic by disregarding the extra propagation caused by NLoS propagation when a direct path does not exist. For microcells, the Manhattan street microcell deployment in Fig. 5.5 is assumed.

4.1.1 EFFECT OF STANDARD DEVIATION OF SHADOWING, σ_S

Fig. 5.6 shows the effect of the shadow standard deviation on the mean and standard deviation of the range estimation error with an early-late discriminator offset $\Delta = 1/2$ and a chip-energy-to-noise ratio, $E_c/N_o = 10$ dB. The mean ranging error increases by approximately 10 m as σ_s increases from 4 to 12 dB. The standard deviation of the ranging error also increases due to the increased variability of the shadowing process.

4.1.2 EFFECT OF E_C/N_O

Fig. 5.7 shows the effect of E_c/N_o on the ranging error with $\Delta = 1/2$ and $\sigma_s = 6$ dB. The increase in ranging error for decreasing E_c/N_o is

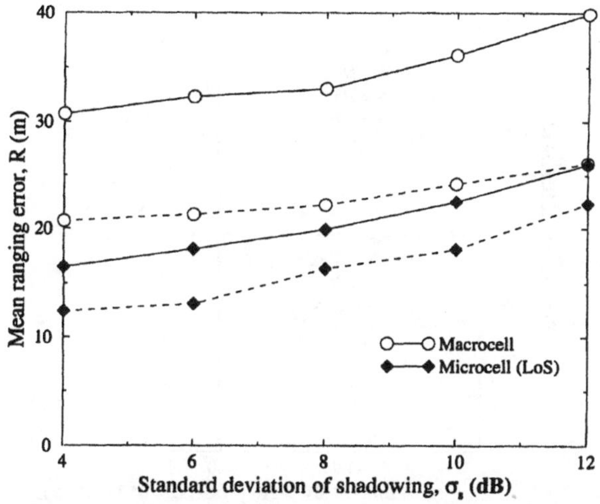

Figure 5.6. Effect of the shadow standard deviation with E_c/N_o=10 dB on the mean (solid lines) and standard deviation (dashed lines) of the ranging error.

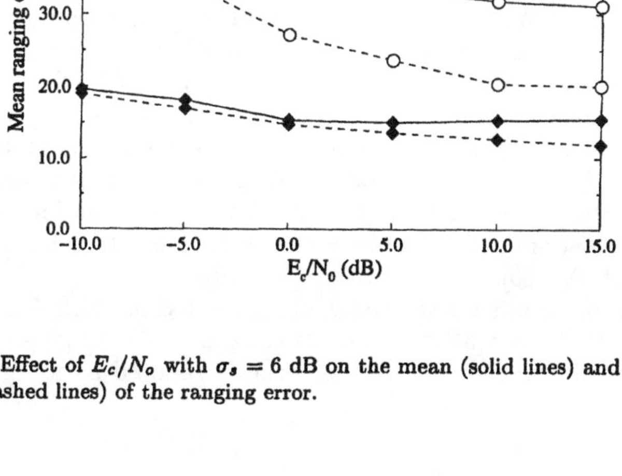

Figure 5.7. Effect of E_c/N_o with $\sigma_s = 6$ dB on the mean (solid lines) and standard deviation (dashed lines) of the ranging error.

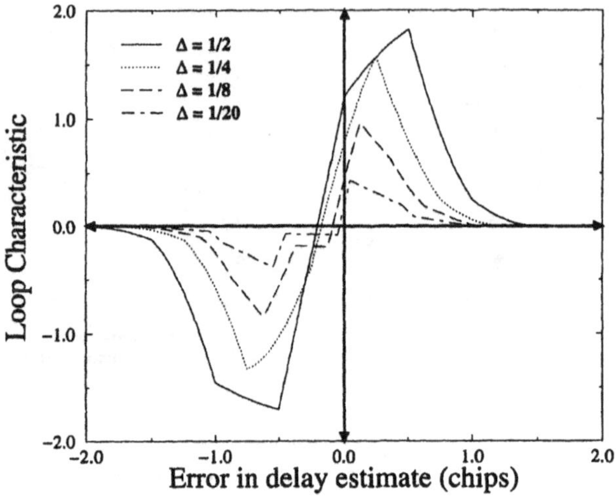

Figure 5.8. Distortion of the S-curve due to multipath for different values of Δ.

expected in any system. The effect is not as pronounced in the micro-cellular environment due to the smaller delay spreads.

4.1.3 EFFECT OF Δ

The effect of multipath on the tracking ability of the DLL can be explained by observing the distortion that multipath causes on the correlation function of the spreading code, which has a triangular shape for a rectangular chip shaping pulse. Fig. 5.8 shows an example of a distorted loop S-curve for the case of two multipath components, the second having half the power of the first and delayed by $T_c/2$. Observe that the tracking error introduced by multipath propagation is reduced by using smaller Δ. However, the minimum size of Δ is limited by hardware considerations (such as the clock rate) and the precorrelation bandwidth in the DLL. Band limiting tends to round the autocorrelation peak which limits the discrimination between the early and late correlation when using small Δ [183].

Simulation results for various Δ are presented in Table 5.2 with $\sigma_s =$ 6 dB and $E_c/N_o = 5$ dB. The results show that the ranging error mean and standard deviation can be significantly reduced by using a smaller Δ.

Table 5.2. Mean ranging error and standard deviation for various values of Δ. Values are in meters.

	Microcell		*Macrocell*	
Δ	*Mean*	*Std. Dev*	*Mean*	*Std. Dev*
1/2	15.6	15.5	30.8	23.6
1/4	10.8	12.1	25.3	19.1
1/8	4.5	6.1	16.9	18.9

4.2 TWO-DIMENSIONAL LOCATION

Two-dimensional location estimates the MS location by using several BSs. Here, we focus on the accuracy of the location estimates as a function of the number of BSs used. This is an important consideration, since using more BSs means more processing and an increased load on the network. Assuming a transmit power of 1 dBW (the maximum for Class III IS-95 MSs), a noise power of -100 dBm was added to each BS receiver for the time-based method. The macrocell and microcell deployment scenarios are as follows.

4.2.1 MACROCELLS

We assume a distance of 6000 m between BSs, i.e., the cell radius is 3000 m as shown in Fig. 5.4. Assuming known BS positions, the MS is randomly placed among the BSs and the nearest BSs are used for the location process. For macrocells, ToA and AoA approaches are compared when using 2, 3, 4, or 5 BSs in the location process as a function of the scattering radius r_s about the MS.

The simulations examined both the unconstrained and constrained location algorithms of Section II-C which almost always converged with $\alpha < 1$ for each BS. The mean and standard deviation of the location error for the ToA method using the unconstrained NL-LS algorithm are shown in Fig. 5.9. For a given scattering radius, the mean and standard deviation decreases when more BSs are used. As expected, a larger scattering radius increases the location error due to NLoS propagation. Recall that LoS propagation is necessary for accurate ranging and location estimates.

The mean and standard deviation of the location error for the ToA method with the constrained NL-LS algorithm are shown in Fig. 5.10. Unlike the unconstrained NL-LS case, the performance is not improved significantly when more BSs are used. Table 5.3 compares the performance of the unconstrained and constrained NL-LS methods, where both

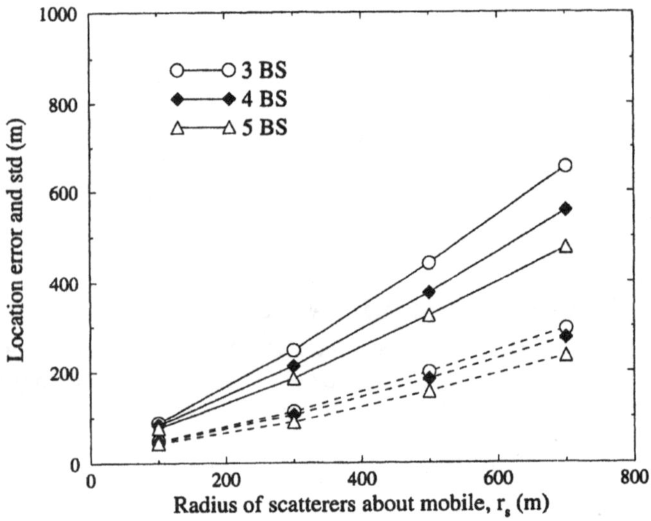

Figure 5.9. Two-dimensional location in macrocells for the ToA method using the unconstrained NL-LS algorithm. Solid lines denote the mean error and dashed lines denote the standard deviation; $\sigma_s = 6$ dB.

Table 5.3. Reduction in the mean location error of the constrained NL-LS algorithm compared to the unconstrained NL-LS algorithm for the ToA method in macrocells. Values are in meters.

# BSs	$r_s = 100$	$r_s = 300$	$r_s = 500$	$r_s = 700$
3	10.0	50.5	125.7	213.2
4	4.8	29.0	75.7	145.0
5	6.1	23.1	53.4	98.8

algorithms are initialized with the same location estimate. The mean location error is reduced up to 30% by using the constrained NL-LS algorithm. The constrained NL-LS with three BSs performs nearly as well as the unconstrained NL-LS with five BSs. This implies that the constrained NL-LS algorithm can result in less network loading.

Results for the AoAradiolocation,AoA method are shown in Fig. 5.11. Once again, a larger number of BSs decreases the mean and standard deviation of the location error. However, diminishing returns are obtained by increasing the number of BSs. For $r_s = 100$ m and $r_s = 300$ m, the

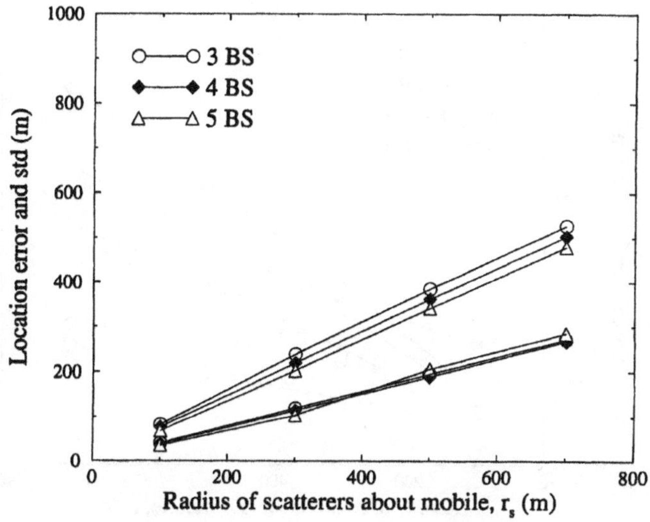

Figure 5.10. Two-dimensional location in macrocells for the ToA method using the constrained NL-LS algorithm. Solid lines denote the mean error and dashed lines denote the standard deviation; $\sigma_s = 6$ dB.

unconstrained ToA method outperforms the AoA method for the same number of BSs. For larger scattering radii, the AoA method steadily improves and performs slightly better than the unconstrained ToA method at a radius of $r_s = 700$ m for the same number of BSs. In all cases, the constrained ToA method performs the best.

4.2.2 MICROCELLS

As the MS rounds the corner from BS0 to BS1 in Fig. 5.5, its location is estimated using a combination of BSs. For this particular deployment, we assume LoS propagation between the MS and two BSs (four at an intersection) and NLoS propagation to the other two BSs. Only the ToA method is used since the AoA method is unreliable due to the relatively large AoA spreads in microcells. The number of BSs used to derive the location estimate ranges from two to four. With 2-BS location, BS0 and BS2 are used until the MS rounds the corner after which BS1 and BS3 are used. With 3-BS location, BS0, BS1 and BS2 are used until the MS rounds the corner after which BS0, BS1 and BS3 are used.

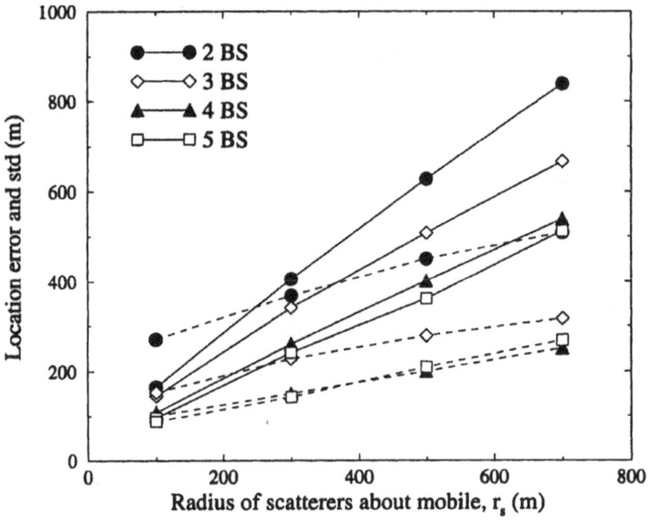

Figure 5.11. **Two-dimensional location in macrocells using the AoA method. Solid lines denote the mean error and dashed lines denote the standard deviation.**

Figs. 5.12 and 5.13 plot the results for the microcell deployment in Fig. 5.5, and indicate that the location accuracy is improved with more BSs for the unconstrained and constrained NL-LS algorithms, respectively. Both figures show the effect of the NL-LS weighting factors, α, on the location performance. The LoS BSs weights are $\alpha = 1$, whereas the NLoS BS weights were varied from 0.2 to 1.0. Fig. 5.12 shows that a smaller α can significantly reduce the mean and standard deviation of the location error with the unconstrained NL-LS algorithm, especially for 3-BS location. For the constrained NL-LS algorithm, there is no significant improvement for $\alpha < 0.6$, as shown in Fig. 5.13. It is interesting to note that 4-BS location is much better than 3-BS location, even though an additional NLoS BS is used. This is because the two NLoS BSs tend to "cancel" one another's effects as a result of the symmetry of the BS layout.

Two-BS location was also considered, using the two LoS BSs with $\alpha = 1$ for both BS. The mean location error was 6.9 m with a standard deviation of 2.0 m for both the unconstrained and constrained NL-LS algorithms. Although 3 BSs are required for 2-D location, the constraint that the MS must lie on a line between the two LoS BSs provides the

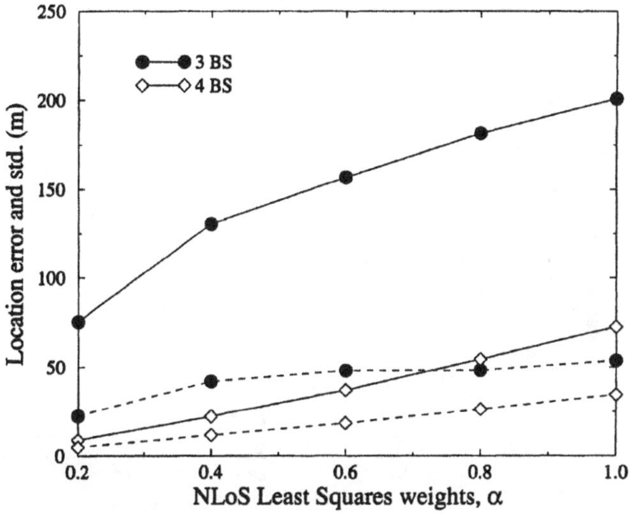

Figure 5.12. Two-dimensional location in microcells using the unconstrained NL-LS algorithm for various values of the NLoS BS weights, α. Solid lines denote the mean error and dashed lines denote the standard deviation.

additional information needed for 2-BS location. This topographical constraint provides more accurate location information than the additional information from a NLoS ToA measurement and, therefore, 2-BS location performs better than 3- or 4-BS location. Note that the feasibility of this approach depends on the BS topography.

5. SUMMARY

This chapter has looked at a computer simulation of ToA and AoA location systems in realistic macrocellular and microcellular environments. The results of the simulation are dependent on the propagation models and parameters chosen as well as the scattering model that was chosen for macrocells. The implementation of an effective cellular radiolocation system requires a method for identifying the set of BSs to be used for deriving the location estimate. One guiding parameter for the choice of BSs is the production of a good GDOP. Another guiding parameter is to choose as many LoS BSs as possible to limit the influence of NLoS propagation effects. This is especially important for microcells, since the use of NLoS BSs can introduce large errors in the location estimates, as

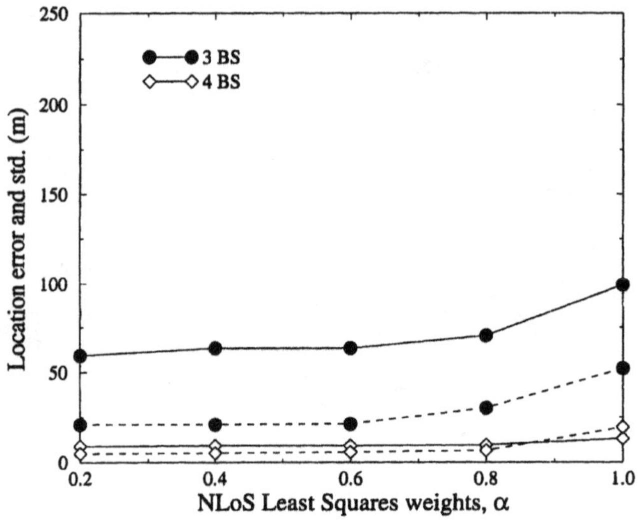

Figure 5.13. Two-dimensional location in microcells using the constrained NL-LS algorithm for various values of the NLoS BS weights, α. Solid lines denote the mean error and dashed lines denote the standard deviation.

it does in macrocells. It may be possible to use the link quality measurements that are used for handoff initiation to determine the appropriate set of BSs. To provide the performance gain by using weighting to mitigate the NLoS problem, it is necessary to determine which BSs are NLoS with the MS. However, this is still an open area of investigation.

Chapter 6

TOA ESTIMATION IN MULTIPLE-ACCESS INTERFERENCE

The performance results of the previous chapter only assumed a single user in the system. When other users are considered, the multiple-access interference will interfere with the tracking ability of conventional time delay trackers, such as the NC-DLL. If the conventional NC-DLL is used to provide ToA estimates for location, the accuracy of the computed location estimates will be directly affected. Consequently, methods must be developed to mitigate the effects of MAI on code acquisition and tracking.

In this chapter, the effects of MAI on conventional code acquisition and tracking are discussed. The performance of the conventional NC-DLL is derived analytically. Following the analysis, we review methods of providing improved ToA estimates for CDMA systems and suggest a new method based on nonlinear filtering.

1. EFFECTS OF MULTIPLE-ACCESS INTERFERENCE

Conventional code acquisition based on correlation methods has been previously investigated in the literature. Madhow and Pursley [104] pointed out that the matched filter acquisition performance in the presence of MAI may impose a limit on the capacity of a CDMA system which is more stringent than one coming from bit error specifications. The acquisition capacity, defined to be the maximum number of simultaneous transmissions supported by the spread spectrum scheme while maintaining acceptable performance, was found to be of the order of $L/\log_{10} L$, where L is the matched filter length. The effects of noise, data modulation and periodicity of the spreading sequences were not analyzed in [104], which also assumed perfect phase recovery. Consequently, Corazza and Esposti [29] generalized the analysis to include these effects in which an acquisition verification procedure was devised to improve performance. Finally, an analysis in [114] showed that the burden of obtaining good performance, relative to the single user case, rests on the choice of spreading code.

The conventional NC-DLL has been extensively used as a code tracking device in spread spectrum communications systems. Non-coherent loops are important because they are relatively insensitive to data modulation and do not presume reliable carrier tracking prior to PN code acquisition as do coherent loops [10]. The majority of performance analyses of the conventional NC-DLL have been for additive white Gaussian noise (AWGN) channels [67, 130], involving both linear and nonlinear approaches. The linear approach is applicable when there is a high SNR, while the nonlinear technique is required for a low SNR [130]. The nonlinear approach uses the renewal process theory developed by Meyr [112]. To improve the performance in AWGN, a few variations of the conventional NC-DLL have been developed [195, 203].

The performance of the conventional NC-DLL is severely degraded in the presence of Doppler shift [10, 176] and multipath fading [99, 151, 184]. Several new code tracking loops have been developed to combat multipath fading and these include the multipath estimating delay lock loop (MEDLL) [185], a "2-mode" DLL [177], and a RAKE DLL [152].

Very little work has been done to examine the effects of MAI on the performance of code tracking using the conventional DLL. Exact analysis of MAI effects is difficult particularly for non-coherent loops since it requires knowledge of the statistics of cross-correlation functions of PN sequences. Recently, Hong *et al* [70] simplified the analysis for the NC-DLL in an asynchronous DS-CDMA system by assuming that the MAI can be approximated as Gaussian noise. However, the Gaussian approximation is not always accurate for short PN codes, when there are few interferers, or when the SNR is high [200]. Also, the approximation is not valid when a single user dominates the multiple-access interference [98].

This section presents a performance analysis of the conventional NC-DLL in the presence of MAI without assuming that the MAI can be approximated as a Gaussian random process. Instead, a simple, but accurate, approximation is used for the discrete normalized aperiodic cross-correlation function (ACF) between two PN sequences in order to simplify the computation of higher order moments in the analysis that follows. The criteria used as the measures of performance are the mean-time-to-lose-lock (MTLL) and the root-mean-square (*rms*) tracking error.

1.1 NON-COHERENT DELAY LOCK LOOPS

We consider the multiple-access system model shown in Fig. 2.4 with K users that is described by equations (2.14) to (2.19). In addition, only a single tap ($M = 1$) is used in the wideband propagation model since we are only interested in analyzing the effects of MAI. The effects of multipath fading (no MAI) on the performance of the NC-DLL has been comprehensively studied in [151]. The form of the low pass filter, $h_{\rm LP}(t)$, will not be specified in order to keep the results of the analysis general. The NC-DLL shown in Fig. 4.4 and described in Chapter 4 is considered.

1.1.1 LOOP DIFFERENTIAL EQUATION

Without loss of generality, we will assume that the signal we desire to track belongs to user 1. Neglecting the code self-noise, the output of the bandpass filter from Fig. 4.4 can be expressed as

$$\begin{aligned} y_{\pm}(t) &= H_{\rm LP}(p)\,[r_{\rm BP}(t)a_1(t-\hat{\tau}_1 \pm \Delta\cdot T_c)] \\ &= \sqrt{2P_1}H_{\rm LP}(p)\,[d_1(t-\tau_1)a_1(t-\tau_1)a_1(t-\hat{\tau}_1 \pm \Delta\cdot T_c)] \\ &\qquad\qquad \cdot\cos(\omega_c t\phi_1) \\ &\quad + \sum_{k=2}^{K}\sqrt{2P_k}H_{\rm LP}(p)[d_k(t-\tau_k)a_k(t-\tau_k)a_1(t-\hat{\tau}_1 \pm \Delta\cdot T_c)] \\ &\qquad\qquad \cdot\cos(\omega_c t+\phi_k) \;+\; n_{\pm}(t) \end{aligned} \tag{6.1}$$

where $H_{\rm LP}(p)$ represents the equivalent lowpass filter in operator form with $p = \partial/\partial t$. The noise terms at the output of the filter are given by

$$\begin{aligned} n_{\pm}(t) &= H_{\rm BP}(p)\,[n_{\rm BP}(t)a_1(t-\hat{\tau}_1 \pm \Delta\cdot T_c)] \\ &= \sqrt{2}n_{\pm}^{c}(t)\cos(\omega_c t) - \sqrt{2}n_{\pm}^{s}(t)\sin(\omega_c t) \end{aligned} \tag{6.2}$$

where $n_{\pm}^{i}(t) = H_{\rm LP}(p)\,[n_i(t)a_1(t-\hat{\tau}_1 \pm \Delta\cdot T_c)]$ for $i = c, s$. The noise terms $n_i(t)$ ($i = c, s$) are defined in (2.19).

Assume that the operation of the VCC is defined by equation (5.17) where $u(t) = F(p)[e(t)] = f(t) * e(t)$ is the output of the loop filter and $*$ denotes convolution. $F(p)$ is the loop filter in operator form. With this definition, the loop bandwidth is very small, allowing only the component of the error at the carrier frequency ω_c (or equivalently, the DC component at baseband) through. As a result, the only component of interest in $y_{\pm}(t)$ is the product of the cosine and the DC component of $d_k(t-\tau_k)a_k(t-\tau_k)a_1(t-\hat{\tau}_1 \pm \Delta T_c)$, for $k = 1, \ldots, K$. Hence, the output of the bandpass arm filters can be expressed as

$$\begin{aligned} y_\pm(t) &= \sqrt{2P_1}\bar{d}_1(t-\tau_1)R_a(\varepsilon \pm \Delta)\cos(\omega_c t + \phi_1) \\ &\quad + \sum_{k=2}^{K} \sqrt{2P_k}\bar{d}_k(t-\tau_k)R_{k,1}(\tau_d^{(k)} + \varepsilon \pm \Delta)\cos(\omega_c t + \phi_k) \\ &\quad + n_\pm(t) \end{aligned} \tag{6.3}$$

where

$$R_{k,i}(x) = \frac{1}{NT_c}\int_0^{NT_c} a_k(t + xT_c)a_i(t)\,dt \tag{6.4}$$

is the *continuous* periodic cross-correlation function between two PN sequences of period NT_c, $\bar{d}_i(t) = H_{\mathrm{LP}}(p)\,[d_i(t)]$ is the filtered data signal, $\varepsilon = (\tau_1 - \hat{\tau}_1)/T_c$ is the normalized tracking error and $\tau_d^{(k)} = (\tau_k - \tau_1)/T_c$ is the normalized delay of the kth user relative to the first and is uniformly distributed on $[0, N)$. As in [151], the autocorrelation function, $R_a(x) \equiv R_{i,i}(x)$ is approximated as

$$R_a(x) = \begin{cases} 1 - |x|, & |x| \le 1 \\ 0, & \text{otherwise} \end{cases} . \tag{6.5}$$

Maximal length sequences (m-sequences) have a two-valued autocorrelation function with values of 1 and $-1/N$. Gold codes, which are popular for CDMA applications, have a four-valued autocorrelation function with values of 1, $-1/N$, $-t(n)/N$, and $(t(n)-2)/N$, where $t(n) = 1 + 2^{\lfloor (n+2)/2 \rfloor}$. In any case, equation (6.5) is a good approximation for the autocorrelation functions of the PN code families of interest.

After squaring and summing the correlator outputs, the error signal which is the input to the loop filter can be expressed as

$$e(t) = P_1\bar{d}_1^2(t-\tau_1)S(\varepsilon, \Delta) + n_T(t, \varepsilon)] \tag{6.6}$$

where

$$S(\varepsilon, \Delta) = R_a^2(\varepsilon - \Delta) - R_a^2(\varepsilon + \Delta) \tag{6.7}$$

$$n_T(t, \varepsilon) = n_1(t) + n_2(t, \varepsilon) + n_3(t, \varepsilon) + n_4(t, \varepsilon) + n_5(t, \varepsilon) \tag{6.8}$$

$$n_1(t) = n_-^2(t) - n_+^2(t) \tag{6.9}$$

$$\begin{aligned} n_2(t, \varepsilon) &= \sqrt{8P_1}\bar{d}_1(t-\tau_1)\,[n_-(t)R_a(\varepsilon - \Delta) - n_+(t)R_a(\varepsilon + \Delta)] \\ &\quad \cdot \cos(\omega_c t + \phi_1) \end{aligned} \tag{6.10}$$

$$n_3(t, \varepsilon) = 2\sum_{k=2}^{K} \sqrt{2P_k}\bar{d}_k(t-\tau_1)\left[n_-(t)R_{k,1}(\tau_d^{(k)} + \varepsilon - \Delta)\right.$$

$$-n_+(t)R_{k,1}(\tau_d^{(k)}+\varepsilon+\Delta)\Big]\cos(\omega_c t+\phi_k) \quad (6.11)$$

$$\begin{aligned} n_4(t,\varepsilon) &= 2\sum_{k=2}^{K}\sqrt{P_1P_k}\bar{d}_1(t-\tau_1)\bar{d}_k(t-\tau_k) \\ &\quad \cdot\Big[R_a(\varepsilon-\Delta)R_{k,1}(\tau_d^{(k)}+\varepsilon-\Delta) \\ &\qquad -R_a(\varepsilon+\Delta)R_{k,1}(\tau_d^{(k)}+\varepsilon+\Delta)\Big] \\ &\qquad \cdot\cos(\phi_k-\phi_1) \end{aligned} \quad (6.12)$$

$$\begin{aligned} n_5(t,\varepsilon) &= \sum_{k=2}^{K}\sum_{j=2}^{K}\sqrt{P_kP_j}\bar{d}_k(t-\tau_k)\bar{d}_j(t-\tau_j) \\ &\quad \cdot\Big[R_{k,1}(\tau_d^{(k)}+\varepsilon-\Delta)R_{j,1}(\tau_d^{(j)}+\varepsilon-\Delta) \\ &\qquad -R_{k,1}(\tau_d^{(k)}+\varepsilon+\Delta)R_{j,1}(\tau_d^{(j)}+\varepsilon+\Delta)\Big] \\ &\qquad \cdot\cos(\phi_k-\phi_j) \end{aligned} \quad (6.13)$$

where $S(\varepsilon,\Delta)$ is the loop S-curve and the $n_i(t,\varepsilon)$ $(i=1,\ldots,5)$ are the equivalent noise terms. The filtering operation causes $\bar{d}_1^2(t-\tau_1)$ to fluctuate around its mean value,

$$D_2 = E[\bar{d}_1^2(t-\tau_1)] = \int_{-\infty}^{\infty} S_d(f)\,|H_{\mathrm{LP}}(f)|^2\,df \quad (6.14)$$

where $S_d(f)$ is the power spectral density (*psd*) of the data sequence and $|H_{\mathrm{LP}}(f)|$ is the magnitude of the equivalent lowpass transfer function for each of the arm filters. The modulation self-noise is considered negligible [130]. Accounting for the above gives the error signal

$$e(t) = P_1D_2S(\varepsilon,\Delta)+n_T(t,\varepsilon)\ . \quad (6.15)$$

The dynamic behavior of the loop is described, in operator form, by the following differential equation which is obtained by combining (5.17) and (6.15):

$$p\varepsilon(t) = p\frac{\tau(t)}{T_c} - K_{\mathrm{vcc}}\cdot F(p)\,[P_1D_2S(\varepsilon,\Delta)+n_T(t,\varepsilon)]\ . \quad (6.16)$$

1.1.2 CROSS-CORRELATION APPROXIMATION

Later in the analysis, it will be important to express $R_{k,i}(x)$ in its discrete form. By letting $x=l+s$, with $l=\lfloor x\rfloor$ and $s=x-l$, $R_{k,i}(x)$ can be expressed as

$$R_{k,i}(x)\equiv R_{k,i}(s,l)=(1-s)\Theta_{k,i}(l)+s\Theta_{k,i}(l+1) \quad (6.17)$$

where $\Theta_{k,i}(l)$ is the discrete periodic cross-correlation function

$$\Theta_{k,i}(l) = \frac{1}{N}\sum_{n=0}^{N-1} a_{k,n}a_{i,n+l} \ . \tag{6.18}$$

For the random variable x uniform on $[0, N)$, the random variables s and l are independent and uniformly distributed on $[0, 1)$ and on the set $\{0, \ldots, N-1\}$, respectively. Additionally, $\Theta_{k,i}(l)$ can be expressed in terms of the discrete normalized ACF, $C_{k,i}(l)$, by $\Theta_{k,i}s(l) = C_{k,i}(l) + C_{k,i}(l-N)$ where [207]

$$C_{k,i}(l) = \begin{cases} \frac{1}{N}\sum_{n=0}^{N-1-l} a_{k,n}a_{i,n+l} & 0 \le l \le N-1 \\ \frac{1}{N}\sum_{n=0}^{N-1+l} a_{k,n-l}a_{i,n} & 1-N \le l \le 0 \\ 0 & \text{else} \end{cases} \ . \tag{6.19}$$

The computation of the multiple access interference terms in (6.10)-(6.13) involves $K-1$ cross-correlation functions which can be computationally expensive if N is large, and is also dependent on the spreading codes chosen for the users. An alternative approach is to treat the $C_{k,i}(l)$ as random variables using a linear model for the discrete *pdf* of the average ACF for a set of PN sequences of length N [23]. Following [23], the *pdf* of the normalized ACF can be approximated by a symmetrical triangle function about the origin as:

$$P[C_{k,i}(l) = \chi] = \begin{cases} \frac{1}{\Lambda^2}[\Lambda - |\chi|], & |\chi| \le \Lambda \\ 0, & \text{otherwise} \end{cases} \tag{6.20}$$

where $\chi \cdot N$ is an integer and $\Lambda = \lfloor\sqrt{3N} - 1\rfloor/N$ and $\lfloor\alpha\rfloor$ denotes the largest integer less than α. The following moments of the random variable $C_{k,i}(l)$ will be required in the analysis that follows:

$$\mathrm{E}[C_{k,i}(l)] = 0 \tag{6.21}$$

$$\mathrm{E}[C_{k,i}^2(l)] = \frac{\Lambda(\Lambda+2)}{6N^2} \tag{6.22}$$

$$\mathrm{E}[C_{k,i}^4(l)] = \frac{\Lambda(\Lambda+2)(2\Lambda^2+4\Lambda-1)}{30N^4} \ . \tag{6.23}$$

1.2 PERFORMANCE ANALYSIS

From the above discussion, the effect of the multiple access interference is to add noise terms to the system. Since the S-curve is not affected through this formulation, both linear and nonlinear analyses will be employed.

1.2.1 LINEAR ANALYSIS

Linear theory is applicable when there is a high loop SNR so that the error ε will be small most of the time causing the DLL to operate in the linear region of $S(\varepsilon, \Delta)$. Consequently, the S-curve can be replaced by its linear equivalent in the vicinity of $\varepsilon = 0$. We note that for high SNR, the MTLL will be quite large so that the tracking error is the only important performance criteria to examine. Assuming a flat power spectral density $S_{n_T}(f, \varepsilon)$ for $n_T(t, \varepsilon)$ within the loop bandwidth, B_L, the tracking error variance is given by [130]

$$\sigma_\varepsilon^2 = \frac{S_{n_T}(0,0)B_L(\Delta)}{[S'(0,\Delta)D_2P_1]^2} \tag{6.24}$$

where $S'(0, \Delta) = \partial S(\varepsilon, \Delta)/ = \partial d\varepsilon|_{\varepsilon=0}$ is the slope of the S-curve at the origin and

$$B_L(\Delta) = 2(1 - \Delta)B_{L_0} \tag{6.25}$$

where B_{L_0} is the nominal closed-loop bandwidth defined as

$$B_{L_0} = B_L(\Delta)|_{\Delta=1/2} = \frac{K_{\text{vcc}}D_2P_1}{2} . \tag{6.26}$$

The term $S_{n_T}(0, 0)$ is [130]

$$S_{n_T}(0,0) = 2\int_{-\infty}^{\infty} R_{n_T}(\xi, \varepsilon)|_{\varepsilon=0}\, d\xi \tag{6.27}$$

where $R_{n_T}(\xi, \varepsilon) = \mathrm{E}[n_T(t, \varepsilon)n_T(t + \xi, \varepsilon)]$ is the autocorrelation function of the total noise, $n_T(t, \varepsilon)$ in (6.8), which is derived in Appendix A.

1.2.2 NONLINEAR ANALYSIS

In the nonlinear theory, the high SNR assumption is removed by using the stochastic differential equation (6.16) to find the stationary *pdf* of the tracking error. For a first-order tracking loop with $F(p) = 1$, Meyr [112] has shown that the stationary *pdf* of the tracking error, $p(\varepsilon)$ satisfies the Fokker-Plank equation

$$\frac{\partial}{\partial\varepsilon}\left[\frac{K_1(\varepsilon)p(\varepsilon)}{1 - P_{LD}(\varepsilon)}\right] - \frac{1}{2}\frac{\partial^2}{\partial\varepsilon^2}\left[\frac{K_2(\varepsilon)p(\varepsilon)}{1 - P_{LD}(\varepsilon)}\right] = \frac{\pi(\varepsilon)}{\mathrm{E}[\tau_{tr}]} \tag{6.28}$$

where $P_{LD}(\varepsilon)$ is the probability that the lock detector will trigger for re-acquisition while the error is still in the permissible range $(\varepsilon_{\min}, \varepsilon_{\max})$, $\pi(\varepsilon)$ is the initial PN code phase error *pdf*, and $\mathrm{E}[\tau_{tr}]$ is the MTLL. The necessary boundary conditions for (6.28) are $p(\varepsilon_{\min}) = p(\varepsilon_{\max}) = 0$.

Additionally, the terms $K_1(\varepsilon)$ and $K_2(\varepsilon)$ are defined as follows [112]:

$$K_2(\varepsilon) = K_{\text{vcc}}^2 \int_{-\infty}^{\infty} R_{n_T}(\xi, \varepsilon)\, d\xi \tag{6.29}$$

$$K_1(\varepsilon) = \frac{1}{T_c}\frac{\partial}{\partial t}\tau(t) - K_{\text{vcc}} P_1 D_2 S(\varepsilon, \Delta) + \frac{1}{4}\frac{\partial}{\partial \varepsilon} K_2(\varepsilon)\ . \tag{6.30}$$

The MTLL and *rms* tracking error can be obtained from the *pdf* of the tracking error, $p(\varepsilon)$ as follows [130]:

$$\mathrm{E}[\tau_{tr}] = \int_{-\infty}^{\infty} Q(x)\, dx \tag{6.31}$$

$$\sigma_{rms} = \sqrt{\int_{\varepsilon_{\min}}^{\varepsilon_{\max}} \varepsilon^2 p(\varepsilon)\, d\varepsilon} \tag{6.32}$$

where $p(\varepsilon)$ is defined by [112]

$$p(\varepsilon) = \frac{Q(\varepsilon)}{\mathrm{E}[\tau_{tr}]} \tag{6.33}$$

and

$$Q(\varepsilon) = \frac{2[1 - P_{LD}(\varepsilon)]e^{-u_o(\varepsilon)}}{K_2(\varepsilon)} \int_{\varepsilon_{\min}}^{\varepsilon} [C_0 - \Pi(z)]e^{u_o(z)}\, dz \tag{6.34}$$

where $\varepsilon_{\min} \leq \varepsilon \leq \varepsilon_{\max}$ and

$$u_o(\varepsilon) = -2\int_{\varepsilon_{\min}}^{\varepsilon} \frac{K_1(x)}{K_2(x)}\, dx \tag{6.35}$$

$$C_0 = \frac{\int_{\varepsilon_{\min}}^{\varepsilon_{\max}} \Pi(x) e^{u_o(x)}\, dx}{\int_{\varepsilon_{\min}}^{\varepsilon_{\max}} e^{u_o(x)}\, dx} \tag{6.36}$$

$$\Pi(\varepsilon) = \int_{\varepsilon_{\min}}^{\varepsilon} \pi(z)\, dz \tag{6.37}$$

where $u_o(\varepsilon)$ is called the *potential function* and $\Pi(\varepsilon)$ is the initial cumulative distribution function, *cdf*, of the tracking error.

1.3 EFFECTS OF MULTIPLE ACCESS INTERFERENCE

This section presents the results of an analysis of the effects of MAI on the performance of the NC-DLL. We define γ_d as the bit SNR (i.e., the SNR in the data bandwidth) and γ_{L_0} as the loop SNR for $\Delta = 1/2$. These parameters are given by

$$\gamma_d = \frac{P_1 T_b}{N_0} \tag{6.38}$$

$$\gamma_{Lo} = \frac{P_1}{N_0 B_{Lo}|_{\Delta=\frac{1}{2}}} \tag{6.39}$$

where T_b is the bit duration, and P_1 is the received power of the first user. The parameter ζ_0 is defined as the ratio γ_{Lo}/γ_d which from (6.38) and (6.39) gives

$$\zeta_0 = \frac{P_1}{N_0 B_{Lo}|_{\Delta=\frac{1}{2}}} = \frac{1}{T_b B_{Lo}|_{\Delta=\frac{1}{2}}} \tag{6.40}$$

1.3.1 EQUATIONS FOR APF APPROXIMATION

From the derivation of $R_{n_T}(\xi, \varepsilon)$ in Appendix A along with the definitions above, the functions $K_2(\varepsilon)$ and $K_1(\varepsilon)$ of equations (6.29) and (6.30), respectively, can be expressed as

$$\begin{aligned}
K_2(\varepsilon) = & \frac{1}{D_2\gamma_{Lo}^2 T_b}\Big\{8C_1\left[1 - R_a^2(2\Delta)\right] \\
& +8\gamma_d C_2\Big[R_a^2(\varepsilon-\Delta) + R_a^2(\varepsilon+\Delta) \\
& \qquad -2R_a(2\Delta)R_a(\varepsilon-\Delta)R_a(\varepsilon+\Delta)\Big] \\
& +\left[\sum_{k=2}^{K}\frac{P_k}{P_1}\right]\frac{16}{9N^2}\gamma_d C_2\Lambda(\Lambda+2)v_1(\Delta) \\
& +\left[\sum_{k=2}^{K}P_k\right]\frac{16}{9N^2}\gamma_d C_3\Lambda(\Lambda+2)\Big[R_a^2(\varepsilon-\Delta) + R_a^2(\varepsilon+\Delta) \\
& \qquad -2v_2(\Delta)R_a(\varepsilon-\Delta)R_a(\varepsilon+\Delta)\Big] \\
& +\left[\sum_{k=2}^{K}\frac{P_k^2}{P_1}\right]\frac{8}{225N^4}\gamma_d C_4\Lambda(\Lambda+2)\Big[(32\Lambda^2+16\Lambda-6) \\
& \qquad +w_1(\Delta)(9\Lambda^2+18\Lambda-2)\Big] \\
& +\left[\sum_{k=2}^{K}\sum_{\substack{j=2\\ j\neq k}}^{K}\frac{P_k P_j}{P_1}\right]\frac{16}{81N^4}\gamma_d C_3\Lambda^2(\Lambda+2)^2 w_2(\Delta)\Big\}
\end{aligned} \tag{6.41}$$

$$\begin{aligned}
K_1(\varepsilon) = & \frac{1}{T_c}\frac{\partial\tau(t)}{\partial t} - \frac{2\gamma_d}{T_b\gamma_{Lo}}S(\varepsilon,\Delta) \\
& +\frac{4}{D_2\gamma_{Lo}^2 T_b}\{\gamma_d C_2\,[R_a(\varepsilon-\Delta) \\
& \quad \cdot\{R_a'(\varepsilon-\Delta) - R_a(2\Delta)R_a'(\varepsilon+\Delta)\} \\
& \quad +R_a(\varepsilon+\Delta)\{R_a'(\varepsilon+\Delta) - R_a(2\Delta)R_a'(\varepsilon-\Delta)\}]
\end{aligned}$$

$$
\begin{aligned}
&+\left[\sum_{k=2}^{K} P_k\right] \frac{2}{9N^2}\gamma_d C_3 \Lambda(\Lambda+2) \\
&\cdot \left[R_a(\varepsilon-\Delta)\left\{R_a'(\varepsilon-\Delta) - v_2(\Delta)R_a'(\varepsilon+\Delta)\right\}\right. \\
&\left.\left.+ R_a(\varepsilon+\Delta)\left\{R_a'(\varepsilon+\Delta) - v_2(\Delta)R_a'(\varepsilon-\Delta)\right\}\right]\right\} \qquad (6.42)
\end{aligned}
$$

where $R_a'(\varepsilon \pm \Delta)$ is the derivative of $R_a(\varepsilon \pm \Delta)$ with respect to ε given by

$$
\begin{aligned}
R_a'(\varepsilon \pm \Delta) &\equiv \frac{\partial}{\partial \varepsilon} R_a(\varepsilon \pm \Delta) \\
&= \begin{cases} 1 & -1 \mp \Delta \le \varepsilon \le \mp\Delta \\ -1 & \mp\Delta \le \varepsilon \le 1 \mp \Delta \end{cases} \qquad (6.43)
\end{aligned}
$$

and the constants, $C_i, i = 1, \ldots, 4$ are defined by

$$C_1 = \frac{T_b}{D_2}\int_{-\infty}^{\infty} |H_{\mathrm{LP}}(f)|^4 \, df \qquad (6.44)$$

$$C_2 = \frac{1}{D_2}\int_{-\infty}^{\infty} S_d(f)|H_{\mathrm{LP}}(f)|^4 \, df \qquad (6.45)$$

$$C_3 = \frac{1}{D_2 N_0}\int_{-\infty}^{\infty} R_{\bar{d}}^2(\xi)\, d\xi \qquad (6.46)$$

$$C_4 = \frac{1}{D_2 N_0}\int_{-\infty}^{\infty} R_{\bar{d}^2}(\xi)\, d\xi \qquad (6.47)$$

where the autocorrelation functions are defined by $R_{\bar{d}}(\varepsilon) = E[\bar{d}(t)\bar{d}(t+\varepsilon)]$ and $R_{\bar{d}^2}(\varepsilon) = E[\bar{d}^2(t)\bar{d}^2(t+\varepsilon)]$. The functions $v_1(\Delta)$, $v_2(\Delta)$, $w_1(\Delta)$, and $w_2(\Delta)$ are defined in Appendix A and we have considered the Doppler shift $\partial\tau(t)/\partial t$ to be negligible.

For the linear analysis, we do not provide the results explicitly. However, we note that $S_{n_T}(0,0)$ is of the same form as $K_2(\varepsilon)$ since $S_{n_T}(0,0) = \frac{2}{K_{\mathrm{vcc}}^2} K_2(\varepsilon)|_{\varepsilon=0}$. Hence, each term $R_a(\varepsilon \pm \Delta)$ can be replaced by $R_a(\Delta)$ since the autocorrelation function is symmetric about the origin.

1.3.2 EQUATIONS FOR GAUSSIAN APPROXIMATION

Results for approximating the MAI as Gaussian noise were derived in [70]. Following the approach in that paper and making a few corrections, the function $K_2(\varepsilon)$ can be shown to be, for general Δ,

$$
\begin{aligned}
K_2(\varepsilon) &= \frac{1}{D_2 \gamma_{Lo}^2 T_b}\left\{8C_1\left[1 - R_a^2(2\Delta)\right]\right. \\
&\quad +8\gamma_d C_2 \left[R_a^2(\varepsilon-\Delta) + R_a^2(\varepsilon+\Delta)\right.
\end{aligned}
$$

$$
\begin{aligned}
&\qquad -2R_a(2\Delta)R_a(\varepsilon-\Delta)R_a(\varepsilon+\Delta)] \\
&+\frac{8}{3}\gamma_d C_1\left[\sum_{k=2}^{K}\frac{P_k}{NP_1}\right][4-R_a(2\Delta)] \\
&+\frac{8}{3}\gamma_d^2 C_2\left[\sum_{k=2}^{K}\frac{P_k}{NP_1}\right]\left\{2\left[R_a^2(\varepsilon-\Delta)+R_a^2(\varepsilon+\Delta)\right]\right. \\
&\qquad \left.-R_a(\varepsilon-\Delta)R_a(\varepsilon+\Delta)\right\} \\
&\left.+\frac{20}{3}\gamma_d^2 C_1\left[\sum_{k=2}^{K}\frac{P_k}{NP_1}\right]^2\right\}. \qquad (6.48)
\end{aligned}
$$

The function $K_1(\varepsilon)$ can be formed using equations (6.48) and (6.30). The results for the linear analysis will not be given for the Gaussian approximation.

1.4 NUMERICAL RESULTS

We consider a system employing NRZ binary data modulation and ideal bandpass arm filters with a one-sided bandwidth of $1/T_b$ Hz. Using these assumptions, it can be shown that $D_2 = 0.902$, $C_1 = 2/D_2$, and $C_2 = 1$ [130]. The constants C_3 and C_4 are difficult to evaluate, so numerical integration was employed. They were found to be approximately $C_3 = 0.65T_b/D_2N_0$ and $C_4 = 0.814/D_2N_0$. Finally, we assume that the *pdf* of the initial code phase error is $\pi(\varepsilon) = \delta(\varepsilon)$ and that

$$
P_{LD}(\varepsilon) = \begin{cases} \frac{4}{9}\varepsilon^2 & |\varepsilon| \le 1.5 \\ 0 & \text{otherwise} \end{cases} \qquad (6.49)
$$

so that the probability of triggering re-acquisition increases as ε increases. Also, note that boundary absorption occurs at $\varepsilon_{\max} = -\varepsilon_{\min} = 1.5$. All of the results assume that the signals of all users are received with equal strength (i.e., perfect power control) unless specifically noted.

The effect of the number of interfering users in a system on the MTLL and *rms* tracking error of the conventional NC-DLL are shown in Figs. 6.1 and 6.2 for different bit SNRs, γ_d, and a PN code sequence length of $N = 127$, $\Delta = 0.5$, and $\zeta_0 = 100$. The black and white symbols correspond to the results given by the ACF and Gaussian approximations, respectively, as will be the case for the remaining figures. The figure shows that as the number of interferers is increased, the MTLL is drastically reduced for both the ACF and Gaussian approximations. For small γ_d (less than 0 dB), the curves are relatively flat indicating that the thermal noise, and not MAI, is dominating the interference. In these situations, the ACF and Gaussian approximations provide the same results, indicating that the Gaussian approximation is accurate for small

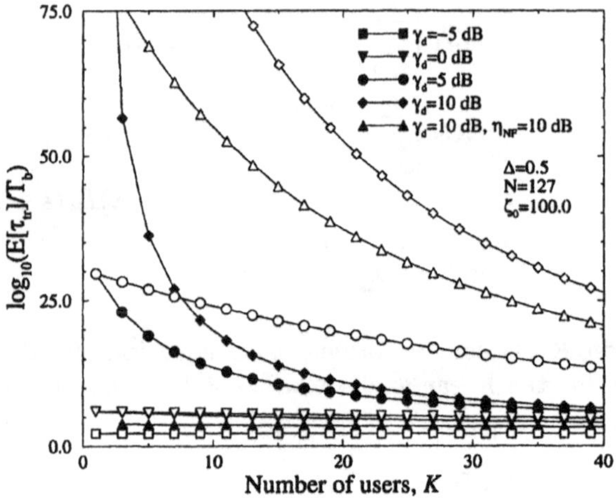

Figure 6.1. The effect of the number of users, K, on the MTLL for different values of γ_d. The black symbols indicate the ACF approximation, while the white symbols indicate the Gaussian approximation.

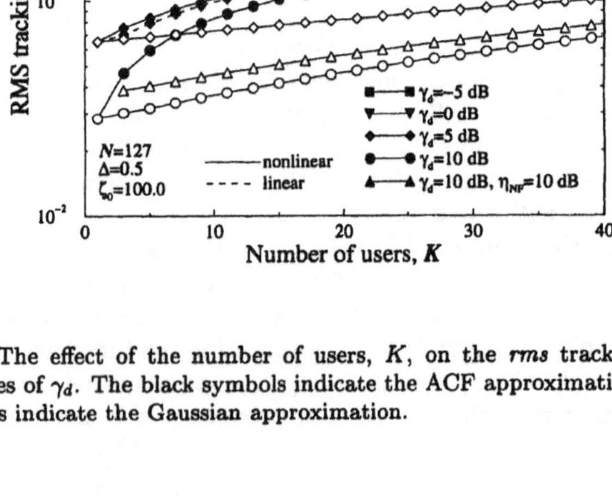

Figure 6.2. The effect of the number of users, K, on the *rms* tracking error for different values of γ_d. The black symbols indicate the ACF approximation, while the white symbols indicate the Gaussian approximation.

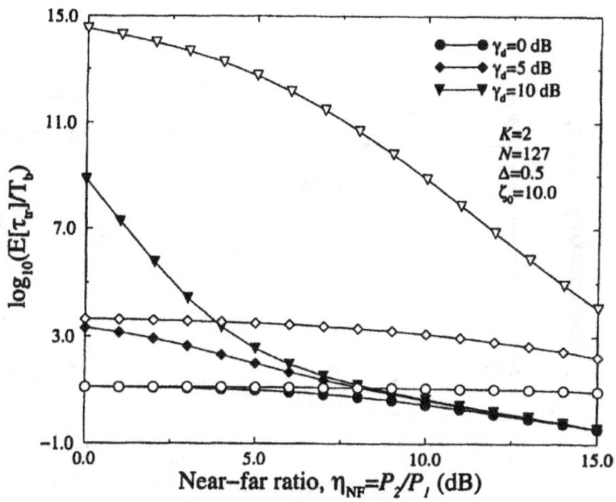

Figure 6.3. The effect of the near-far ratio, $\eta_{\rm NF} = P_2/P_1$, on the MTLL in a two-user system for various values of γ_d. The black symbols indicate the ACF approximation and the white symbols indicate the Gaussian approximation.

γ_d. For larger γ_d (greater than 5 dB), the curves decrease with increasing K, indicating that MAI is dominating performance. Also, for $\gamma_d > 5$ dB, the Gaussian approximation greatly overestimates the the performance of the NC-DLL. The figures also show the effect of a single interferer having power 10 dB above the desired user ($\eta_{\rm NF} = 10$ dB). In this case, the performance is very poor, even though $\gamma_d = 10$ dB. Again, the Gaussian approximation greatly overestimates the performance. The effect of the Gaussian approximation for the MAI on the analysis is to simply increase the level of the Gaussian noise at the receiver, as the curves indicate. Fig. 6.2 also provides both the linear and nonlinear analyses for the *rms* tracking error for the ACF approximation only. The linear analysis proves to be very accurate for high γ_d (5 dB and greater).

In Figs. 6.3 and 6.4, a two user system is considered to examine the effect of the near-far ratio, $\eta_{\rm NF} = P_2/P_1$, on the performance of the NC-DLL. The figures show the MTLL and *rms* tracking error as a function of $\eta_{\rm NF}$ for different γ_d with $N = 127$, $\Delta = 0.5$, and $\zeta_0 = 10.0$. As expected, the performance suffers in a near-far environment. When γ_d is small, the performance remains relatively the same regardless of the near-far ratio, again indicating that the system is dominated by thermal noise. In addition, increasing γ_d provides no improvement in performance for large

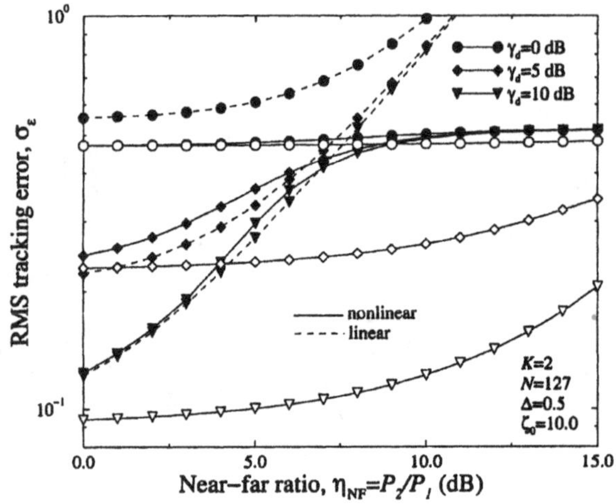

Figure 6.4. The effect of the near-far ratio, $\eta_{\rm NF} = P_2/P_1$, on the *rms* tracking error in a two-user system for various values of γ_d. The black symbols indicate the ACF approximation and the white symbols indicate the Gaussian approximation.

values of the near-far ratio. The figure also indicates that the Gaussian approximation is accurate for low SNRs, but greatly overestimates the performance for high SNRs. Linear and nonlinear analyses are shown for the ACF approximation in Fig. 6.4. The accuracy of the linear results is found to greatly diminish as the near-far ratio is increased. Likewise, the linear results are less accurate for low SNRs.

The effect of the PN code sequence length on the MTLL and *rms* tracking error is shown in Figs. 6.5 and 6.6 for different numbers of users in the system and $\gamma_d = 5$ dB, $\Delta = 0.5$, and $\zeta_0 = 100.0$. For large values of N, the normalized cross-correlation between two users' codes is reduced. Consequently, the performance of the NC-DLL improves as N increases. In fact, the performance of a K user system approaches that of a single user system ($K = 1$) for N large enough. Also, notice that the accuracy of the Gaussian approximation improves with increasing N. While the accuracy of the Gaussian approximation is poor for sequence lengths less than 1000, it is very accurate for lengths greater than 1000 for all the values of K shown. The linear results of Fig. 6.6 for the ACF approximation also prove to be accurate for larger values of N ($N > 100$).

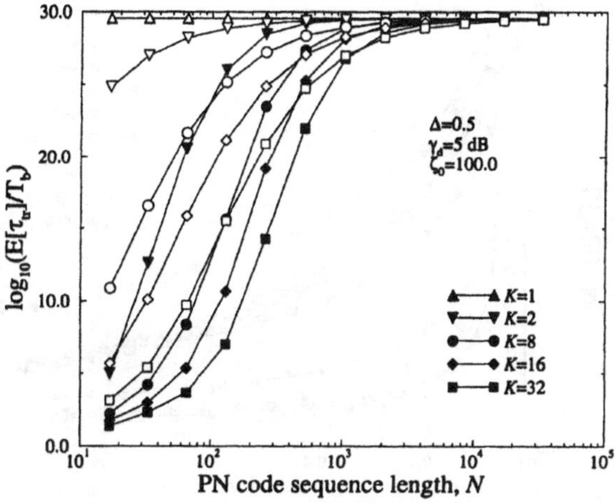

Figure 6.5. The effect of PN code sequence length, N, on the MTLL for various numbers of users. The black symbols indicate the ACF approximation and the white symbols indicate the Gaussian approximation.

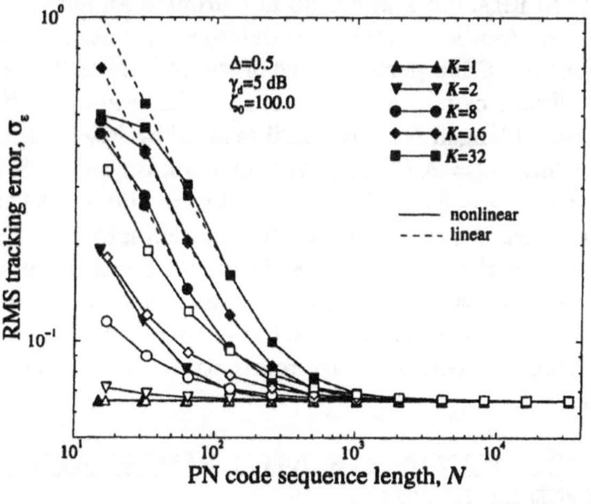

Figure 6.6. The effect of PN code sequence length, N, on the *rms* tracking error for various numbers of users. The black symbols indicate the ACF approximation and the white symbols indicate the Gaussian approximation.

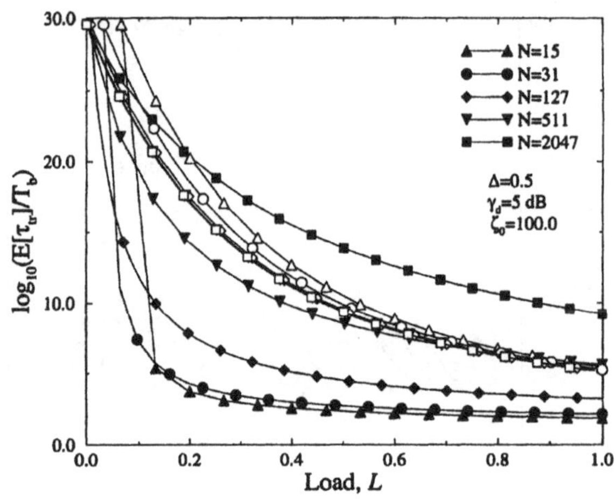

Figure 6.7. The effect of the load, $\mathcal{L} = K/N$, on the MTLL for various PN code sequence lengths, N. The black symbols indicate the ACF approximation and the white symbols indicate the Gaussian approximation.

The results in Figs. 6.5 and 6.6 do not provide an accurate indication of the effect of system load, which we define as the number of users for a given code length, on the performance of the NC-DLL. The code lengths shown in the figure can support more users than shown. For instance, with Gold codes of length N, it is possible to allow $N+2$ users. The effect of the system load $\mathcal{L} = K/N$ is shown in Figs. 6.7 and 6.8 for different code lengths with $\gamma_d = 5$ dB, $\Delta = 0.5$, and $\zeta_0 = 100.0$. As shown in the figure, the performance decreases as the load is increased. Furthermore, the performance of the shorter codes decreases much more quickly than the longer codes as load is increased. Again, this can be attributed to the reduced normalized cross-correlation between users' code sequences. The results from the Gaussian approximation show little difference as N is increased.

2. MITIGATION OF MULTIPLE-ACCESS INTERFERENCE

As shown in the previous section, the conventional receiver, which is based on correlation techniques for data detection and parameter estimation (if necessary), fails in a near-far environment. It is the non-orthogonality of the PN codes of the different users which gives rise to

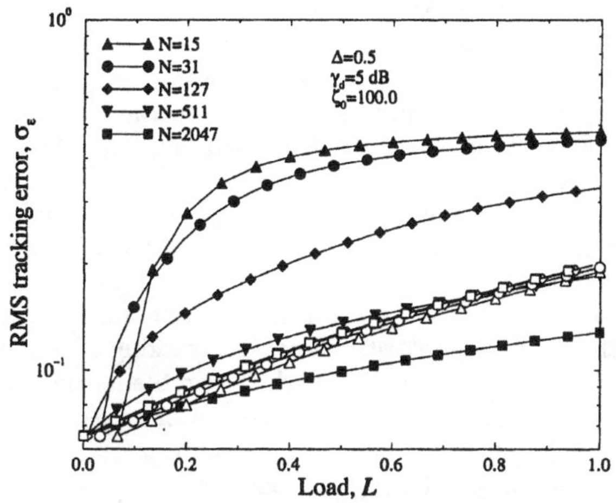

Figure 6.8. The effect of the load, $\mathcal{L} = K/N$, on the *rms* tracking error for various PN code sequence lengths, N. The black symbols indicate the ACF approximation and the white symbols indicate the Gaussian approximation.

the interference. Even with perfect power control, the performance is degraded. Better performance may be possible if the receiver incorporates information about the users' waveforms. This observation has led to the development of near-far resistant detectors and parameter estimators.

Due to the complexity of the optimal receiver developed by Verdú [188, 189], researchers have looked for receivers that provide near-optimal performance, but with much less complexity. These include linear multiuser detectors [34, 101, 187, 199] and subtractive interference cancellers [68, 201]. Many of the proposed detectors assume perfect knowledge of the channel amplitudes, the propagation delays, and/or the carrier phases which, in practice, need to be estimated. Inaccurate estimates will degrade performance of the multiuser detectors, although the detectors will still perform better than their conventional counterparts [172, 206].

The need for accurate parameter estimates in the presence of MAI have spawned the development of joint multiuser detectors/parameter estimators [78, 133]. In addition, near-far resistant parameter estimators for constant or slowly varying channel amplitudes and delays have been developed based on signal subspace techniques [8, 174]. These have been extended to the case of multipath channels with constant channel taps and delays in [7]. An extension to time-varying channel amplitudes with constant or slowly varying time-delays was developed in [173]. An method that extended the subspace methods to estimation of time-varying delays was given in [81]. The development of code acquisition using interference cancellation was presented in [153], while code tracking of time-varying delays based on the DLL combined with MAI cancellation have also been developed [89, 205]. From the results of Chapter 5 and the results of the previous section, we find that near-far resistant parameter estimators are applicable not only for CDMA detection, but can also be useful for time-based radiolocation in CDMA cellular networks.

In the following section, we extend the development of multiuser parameter estimators for joint estimation and tracking of time-varying amplitudes and time delays. We approach the problem from a system identification point of view where an adaptive filter is used to approximate the output of the channel. Due to the nonlinear nature of time delay estimation, a nonlinear filtering approach to estimation was used. The unscented filter (UF) by Julier *et al* [82, 83] is employed which is an extension of the standard Kalman filter (KF) to nonlinear systems and does not approximate the nonlinear estimation problem with a linear one as does the extended Kalman filter (EKF). Thus, there is no need to calculate Jacobian matrices. The complexity of the algorithm is greatly reduced from that of the standard EKF, making it applicable to estimation of closely sampled waveforms.

2.1 DEVELOPMENT OF THE ESTIMATOR

We consider a CDMA system with K users as illustrated in Fig. 2.4. The channel model considered here does not incorporate multiple taps of the wideband channel model as presented in Chapter 2. Rather, only a single tap channel model ($M = 1$) is considered. However, the results of this chapter can easily be extended to the multiple tap channel model as done in [76] for a single-user system.

2.1.1 PROBLEM FORMULATION

Given the samples of the received signal from equation (2.24), $r(l)$, we wish to obtain minimum variance estimates of the unknown parameters β_i and τ_i for $i = 1, \ldots, K$ which are given by

$$\hat{\beta}_i(l|l) = \mathrm{E}\left\{\beta_i(l)|\mathcal{R}^l\right\} \tag{6.50}$$

$$\hat{\tau}_i(l|l) = \mathrm{E}\left\{\tau_i(l)|\mathcal{R}^l\right\} \tag{6.51}$$

where $\mathcal{R}^l$ denotes the set of received samples $\{r(l), r(l-1), \ldots, r(0)\}$. Let the unknown parameters be represented by the $2K \times 1$ vector

$$\boldsymbol{\theta}_p = \begin{bmatrix} \boldsymbol{\beta} \\ \boldsymbol{\tau} \end{bmatrix} \tag{6.52}$$

where $\boldsymbol{\beta} = [\beta_1, \ldots, \beta_K]^T$ and $\boldsymbol{\tau} = [\tau_1, \ldots, \tau_K]^T$, and we have dropped the time dependence for notational convenience. The channel amplitudes and propagation delays of the K users are assumed to obey a Gauss-Markov dynamic channel model, i.e.,

$$\boldsymbol{\beta}(l+1) = \boldsymbol{\Phi}_\beta \boldsymbol{\beta}(l) + \boldsymbol{w}_\beta(l) \tag{6.53}$$

$$\boldsymbol{\tau}(l+1) = \boldsymbol{\Phi}_\tau \boldsymbol{\tau}(l) + \boldsymbol{w}_\tau(l) \tag{6.54}$$

where $\boldsymbol{\Phi}_\beta$ and $\boldsymbol{\Phi}_\tau$ are the $K \times K$ state transition matrices for the amplitudes and delays, respectively. The noise vectors, $\boldsymbol{w}_\beta$ and $\boldsymbol{w}_\tau$, are $K \times 1$ mutually independent Gaussian random vectors with zero mean and

$$\mathrm{E}\left\{\boldsymbol{w}_\beta(i)\boldsymbol{w}_\beta^T(j)\right\} = \delta_{ij}\boldsymbol{Q}_\beta \tag{6.55}$$

$$\mathrm{E}\left\{\boldsymbol{w}_\tau(i)\boldsymbol{w}_\tau^T(j)\right\} = \delta_{ij}\boldsymbol{Q}_\tau \tag{6.56}$$

$$\mathrm{E}\left\{\boldsymbol{w}_\beta(i)\boldsymbol{w}_\tau^T(j)\right\} = \mathbf{0}, \ \forall i, j \tag{6.57}$$

where $\boldsymbol{Q}_\beta = \sigma_\beta^2 \boldsymbol{I}$ and $\boldsymbol{Q}_\tau = \sigma_\tau^2 \boldsymbol{I}$ are the $K \times K$ covariance matrices of the process noises $\boldsymbol{w}_\beta$ and $\boldsymbol{w}_\tau$, respectively, and δ_{ij} is the Kronecker delta function which is unity when $i = j$, and is zero otherwise. We note that the form of $\beta_k(l)$ in (6.53) corresponds to a Rayleigh uncorrelated scattering model for the channel [76].

Using equations (6.52) through (6.54) we can write the state model as

$$\boldsymbol{\theta}_p(l+1) = \boldsymbol{\Phi}\boldsymbol{\theta}_p(l) + \boldsymbol{w}(l) \tag{6.58}$$

where

$$\boldsymbol{\Phi} = \begin{bmatrix} \boldsymbol{\Phi}_\beta & \mathbf{0} \\ \mathbf{0} & \boldsymbol{\Phi}_\tau \end{bmatrix} \tag{6.59}$$

is the $2K \times 2K$ state transition matrix and $\boldsymbol{w}^T(l) = [\boldsymbol{w}_\beta^T(l) \ \boldsymbol{w}_\tau^T(l)]$ is the $2K \times 1$ process noise vector with zero mean and covariance matrix

$$\boldsymbol{Q} = \begin{bmatrix} \boldsymbol{Q}_\beta & \mathbf{0} \\ \mathbf{0} & \boldsymbol{Q}_\tau \end{bmatrix}. \tag{6.60}$$

The scalar measurement model follows from the sampled received signal of (2.24) by

$$z(l) = h\left(\boldsymbol{\theta}_p(l)\right) + v(l) \tag{6.61}$$

where, from (2.24) with $M = 1$, the measurement $z(l) = r(l)$, $v(l) = n(l)$, and

$$h(\boldsymbol{\theta}_p(l)) = \sum_{k=1}^{K} \beta_k(l) d_{k,m_l} \bar{a}_k(lT_s - mT_b - \tau_k(l)) \tag{6.62}$$

Hence, our goal is to find the $2K \times 1$ joint estimator

$$\hat{\boldsymbol{\theta}}_p(l|l) = \mathrm{E}\left\{\boldsymbol{\theta}_p(l)|\mathcal{R}^l\right\} \tag{6.63}$$

with estimated error covariance

$$\mathcal{P} = \mathrm{E}\left\{[\boldsymbol{\theta}_p(l) - \hat{\boldsymbol{\theta}}_p(l|l)][\boldsymbol{\theta}_p(l) - \hat{\boldsymbol{\theta}}_p(l|l)]^T \mid \mathcal{R}^l\right\} . \tag{6.64}$$

A well-known means of forming the minimum variance estimate is the Kalman Filter (KF) [2], which is given in Appendix B. When the state and measurement models are linear in the unknown parameters, the traditional Kalman filter provides the minimum variance estimator. However, the application of the KF to nonlinear systems can be difficult. The most common approach is to use the EKF which simply linearizes all nonlinear models so that the traditional KF can be applied. The EKF is based on the assumption that the models are locally linear, and if they are not, the filter can diverge. Additionally, the linearization process requires the derivation of Jacobian matrices. In the next section, we will explore a nonlinear filtering alternative to the traditional EKF, known as the *unscented filter*, which overcomes these two main difficulties.

2.1.2 THE UNSCENTED FILTER

The optimal nonlinear filter requires that a complete description of the conditional probability density $p(\boldsymbol{\theta}_p(l)|\mathcal{R}^l)$ be maintained and propagated in time. Unfortunately, the exact description of the density requires a potentially unbounded number of parameters, such as moments. A good approximation to this approach, but with greatly decreased complexity, is the UF developed by Julier and Uhlmann [83]. The UF yields performance equivalent to the KF for linear systems, and generalizes nicely to nonlinear systems without the linearization steps required by the EKF. The UF update equations follow from the standard KF which is given in Appendix B. However, the computations of the means and covariances are changed to allow their proper transformation through the nonlinear equations, as we now describe.

The fundamental component of the filter is the *unscented transformation* which provides a method for calculating the statistics of a random variable which undergoes a nonlinear transformation [82]. The transformation uses a set of appropriately weighted points to parameterize the mean and covariance information while permitting the direct propagation of that information through a set of nonlinear equations. The set of points (called *sigma points*) are chosen such that their sample mean and sample covariance are $\hat{\boldsymbol{\theta}}_p(l|l)$ and $\boldsymbol{P}(l|l)$, respectively. For an m-dimensional state space, a set of $2m$ points is necessary and sufficient to form a discrete distribution which is symmetric and has the same mean and covariance as a given state estimate [132]. The UF update equations for our problem are briefly described below.

The set of $4K$ sigma points, $\boldsymbol{\sigma}(l|l)$, is computed from the $2K \times 2K$ matrix $\boldsymbol{P}(l|l)$ by taking its columns from $\pm\sqrt{(2K+\kappa)(\boldsymbol{P}(l|l)+\boldsymbol{Q})}$. The choice of the parameter κ will be discussed at the end of this section. The translated sigma points are then formed from

$$\boldsymbol{\Theta}_0(l|l) = \hat{\boldsymbol{\theta}}_p(l|l) \tag{6.65}$$

$$\boldsymbol{\Theta}_i(l|l) = \boldsymbol{\sigma}_i(l|l) + \hat{\boldsymbol{\theta}}_p(l|l) \tag{6.66}$$

where $\boldsymbol{\sigma}_i(l|l)$ is the ith column of $\boldsymbol{\sigma}(l|l)$. This provides us with points which completely describe the mean, $\hat{\boldsymbol{\theta}}_p(l|l)$, and covariance, $\boldsymbol{P}(l|l)$. The translated points are transformed using the state transition matrix:

$$\boldsymbol{\Theta}_i(l+1|l) = \boldsymbol{\Phi}\boldsymbol{\Theta}_i(l|l) \ . \tag{6.67}$$

The predicted mean $\hat{\boldsymbol{\theta}}_p(l+1|l)$ and the predicted error covariance are then found by

$$\hat{\boldsymbol{\theta}}_p(l+1|l) = \sum_{i=0}^{4K} W_i \boldsymbol{\Theta}_i(l+1|l) \tag{6.68}$$

$$\boldsymbol{P}(l+1|l) = \sum_{i=0}^{4K} W_i \left[\boldsymbol{\Theta}_i(l+1|l) - \hat{\boldsymbol{\theta}}_p(l+1|l)\right] \cdot \left[\boldsymbol{\Theta}_i(l+1|l) - \hat{\boldsymbol{\theta}}_p(l+1|l)\right]^T \tag{6.69}$$

where $W_0 = \kappa/(2K+\kappa)$ and $W_i = 1/2(2K+\kappa)$, for $i = 1, \ldots, 4K$. By passing the translated sigma points through the nonlinear observation function $h(\boldsymbol{\theta}_p(l))$:

$$\mathcal{Z}_i(l+1|l) = h\left(\boldsymbol{\Theta}_i(l+1|l)\right) \tag{6.70}$$

from which we are able to form the predicted estimate

$$\hat{z}(l+1|l) = \sum_{i=0}^{4K} W_i \mathcal{Z}_i(l+1|l) \ . \tag{6.71}$$

From (B.9), we see that the form of the innovation covariance is given by

$$P_{\nu\nu}(l+1|l) = \sigma_n^2 + \sum_{i=0}^{4K} W_i \left[\mathcal{Z}_i(l+1|l) - \hat{z}(l+1|l)\right] \quad (6.72)$$

$$\cdot \left[\mathcal{Z}_i(l+1|l) - \hat{z}(l+1|l)\right]^T \; . \quad (6.73)$$

With the formation of the cross-covariance matrix of $\boldsymbol{\theta}_p$ and z, which can be found by

$$P_{\theta z}(l+1|l) = \sum_{i=0}^{4K} W_i \left[\boldsymbol{\Theta}_i(l+1|l) - \hat{\boldsymbol{\theta}}_p(l+1|l)\right] \quad (6.74)$$

$$\cdot \left[\mathcal{Z}_i(l+1|l) - \hat{z}(l+1|l)\right]^T \; , \quad (6.75)$$

the Kalman gain matrix, $\boldsymbol{\Gamma}_{\mathrm{KF}}(l+1)$, can be found according to (B.8). Finally, the parameter estimate and the the error covariance matrix are updated according to (B.6) and (B.7), after which the process repeats.

The term κ provides an extra degree of freedom to fine tune the higher order moments of the approximation, and hence, can be used to reduce the overall prediction errors. When $\boldsymbol{\theta}_p(l)$ is assumed to be Gaussian, a useful choice for this quantity is $\kappa = 3 - m$, where m is the dimension of the state space, which minimizes the mean squared error up to fourth-order [83]. Also, κ can be chosen to reflect other probability distributions as well. However, if $\kappa < 0$, as it is in our case for $K \geq 2$, the predicted covariance may be non-positive semidefinite. In this case, it is possible to use a modified form of the algorithm by computing the mean as before, but setting $W_0 = 0$ for the predicted covariances [82].

2.2 CRAMÉR-RAO BOUND

The CRLB provides a lower bound for the covariance matrix of the estimation error for any unbiased estimator [5]. Suppose that $\hat{\boldsymbol{\theta}}_p$ is an unbiased estimator of a vector of deterministic unknown parameters $\boldsymbol{\theta}_p$; hence, $\mathrm{E}[\hat{\boldsymbol{\theta}}_p] = \boldsymbol{\theta}_p$. According to the CRLB, the estimator's covariance satisfies

$$\mathrm{E}\left\{\left[\boldsymbol{\theta}_p - \hat{\boldsymbol{\theta}}_p\right]\left[\boldsymbol{\theta}_p - \hat{\boldsymbol{\theta}}_p\right]^T\right\} \geq \boldsymbol{J}^{-1} \quad (6.76)$$

where $\boldsymbol{J}$ is the $2K \times 2K$ Fisher information matrix given by

$$\boldsymbol{J} = \mathrm{E}\left\{\left[\frac{\partial}{\partial \boldsymbol{\theta}_p} \ln \Lambda(\boldsymbol{r})\right]\left[\frac{\partial}{\partial \boldsymbol{\theta}_p} \ln \Lambda(\boldsymbol{r})\right]^T\right\} \quad (6.77)$$

and $\Lambda(\boldsymbol{r})$ is the likelihood function of the received data vector $\boldsymbol{r}$ with respect to $\boldsymbol{\theta}_p$.

We define a $L \times 1$ vector of received samples as $\boldsymbol{r} = [r(1), \ldots, r(L)]^T$. Using (2.25), we can write

$$\boldsymbol{r} = \boldsymbol{s} + \boldsymbol{n} \tag{6.78}$$

where

$$\boldsymbol{s} = \begin{bmatrix} s(1) \\ \vdots \\ s(L) \end{bmatrix} = \begin{bmatrix} \boldsymbol{\beta}^T(1)\boldsymbol{\mathcal{D}}(1)\bar{\boldsymbol{a}}(1) \\ \vdots \\ \boldsymbol{\beta}^T(L)\boldsymbol{\mathcal{D}}(L)\bar{\boldsymbol{a}}(L) \end{bmatrix} \quad \text{and} \quad \boldsymbol{n} = \begin{bmatrix} n(1) \\ \vdots \\ n(L) \end{bmatrix} \tag{6.79}$$

Since $\boldsymbol{n}$ is a vector of independent Gaussian random variables, we find that the likelihood function of $\boldsymbol{r}$ given $\boldsymbol{\theta}_p$ can be expressed as

$$\Lambda(\boldsymbol{r}) = \frac{1}{(2\pi\sigma_n^2)^{L/2}} \exp\left\{-\frac{1}{2\sigma_n^2}\sum_{l=1}^{L}[r(l) - s(l)]^2\right\} \tag{6.80}$$

from which the log-likelihood function directly follows:

$$\ln \Lambda(\boldsymbol{r}) = C - \frac{1}{2\sigma_n^2}\sum_{l=1}^{L}\left[r(l) - \sum_{k=1}^{K}\beta_k(l)d_{k,m_l}\bar{a}_k(lT_s - m_lT_b - \tau_k(l))\right]^2 \tag{6.81}$$

where C is some constant. From Appendix C, we find that the Fisher information matrix, $\boldsymbol{J}$, can be written as a 2×2 block matrix:

$$\boldsymbol{J} = \begin{bmatrix} \boldsymbol{J}_{\beta\beta} & \boldsymbol{J}_{\beta\tau} \\ \boldsymbol{J}_{\beta\tau}^T & \boldsymbol{J}_{\tau\tau} \end{bmatrix} \tag{6.82}$$

where the $K \times K$ matrices of each block are given by

$$\boldsymbol{J}_{\beta\beta} = \frac{1}{\sigma_n^2}\sum_{l=1}^{L}\boldsymbol{\mathcal{D}}(l)\bar{\boldsymbol{a}}(l)\bar{\boldsymbol{a}}^T(l)\boldsymbol{\mathcal{D}}(l) \tag{6.83}$$

$$\boldsymbol{J}_{\beta\tau} = \frac{1}{\sigma_n^2}\sum_{l=1}^{L}\boldsymbol{\mathcal{D}}(l)\bar{\boldsymbol{a}}(l)\bar{\boldsymbol{a}}_d^T(l)\boldsymbol{\mathcal{D}}(l)\boldsymbol{B}(l) \tag{6.84}$$

$$\boldsymbol{J}_{\tau\tau} = \frac{1}{\sigma_n^2}\sum_{l=1}^{L}\boldsymbol{B}(l)\boldsymbol{\mathcal{D}}(l)\bar{\boldsymbol{a}}_d\bar{\boldsymbol{a}}_d^T\boldsymbol{\mathcal{D}}(l)\boldsymbol{B}(l) \ . \tag{6.85}$$

The $K \times K$ matrix $\boldsymbol{B}(l) = \text{diag}[\boldsymbol{\beta}(l)]$, where diag$[\boldsymbol{x}]$ denotes the diagonal matrix composed of the elements from the vector $\boldsymbol{x}$, and $\bar{\boldsymbol{a}}_d(l)$ is the vector of derivatives defined by

$$\bar{\boldsymbol{a}}_d(l) = \begin{bmatrix} \frac{\partial}{\partial \tau_1}\bar{a}_1(lT_s - m_lT_b - \tau_1(l)) \\ \vdots \\ \frac{\partial}{\partial \tau_K}\bar{a}_K(lT_s - m_lT_b - \tau_K(l)) \end{bmatrix} \tag{6.86}$$

where

$$\begin{aligned} \frac{\partial}{\partial \tau_i}\bar{a}_i(lT_s - m_lT_b - \tau_i(l)) &= \\ & \sum_{n=0}^{N-1} a_{i,n}\frac{1}{\pi}\left[\text{sinc}\left(2\frac{lT_s - nT_c - m_lT_b - \tau_i(l)}{T_c}\right)\right. \\ & \left. -\text{sinc}\left(2\frac{lT_s - (n+1)T_c - m_lT_b - \tau_i(l)}{T_c}\right)\right] \end{aligned} \tag{6.87}$$

and $\text{sinc}(y) = \sin(\pi y)/\pi y$. Consequently, from (6.83)–(6.85), the CRLBs for the estimator of the amplitudes and delays is given by (C.13) and (C.14) in Appendix C:

$$\text{CRLB}(\boldsymbol{\beta}) = \left[\boldsymbol{J}_{\boldsymbol{\beta\beta}} - \boldsymbol{J}_{\boldsymbol{\beta\tau}}\boldsymbol{J}_{\boldsymbol{\tau\tau}}^{-1}\boldsymbol{J}_{\boldsymbol{\beta\tau}}^T\right]^{-1} \tag{6.88}$$

$$\text{CRLB}(\boldsymbol{\tau}) = \left[\boldsymbol{J}_{\boldsymbol{\tau\tau}} - \boldsymbol{J}_{\boldsymbol{\beta\tau}}^T\boldsymbol{J}_{\boldsymbol{\beta\beta}}^{-1}\boldsymbol{J}_{\boldsymbol{\beta\tau}}\right]^{-1}. \tag{6.89}$$

The CRLB of the amplitudes and delays for each user is given by the appropriate diagonal element of each CRLB formula. It is desirable that the performance of the developed estimator/tracker approach or meet the CRLB. Whether the algorithm based on the UF approach meets the CRLB will be shown in the results of the next section.

2.3 SIMULATION RESULTS

We now examine the performance of the UF for making parameter estimates for a multiuser detector. We simulate a two-user scenario where the users' PN spreading codes are chosen from the set of Gold codes of length 31 and generated by the polynomials $x^5 + x^2 + 1$ and $x^5 + x^4 + x^3 + x^2 + 1$. The SNR (signal-to-thermal noise ratio) at the receiver of the weaker user was 10 dB. For the state model, the augmented state transition matrix of equation (6.60) was chosen to be $\boldsymbol{\Phi} = 0.999\boldsymbol{I}_{2K\times 2K}$. Also, the process noise covariance matrix was $\boldsymbol{Q} = 0.001\boldsymbol{I}_{2K\times 2K}$. Unless otherwise noted, the SNR for the weaker user is set at 10 dB and the oversampling factor is $q = 2$. One aspect about using Kalman filters or the unscented filter is that they require proper initialization. Depending on the problem, the initial guesses may need to be close to the correct

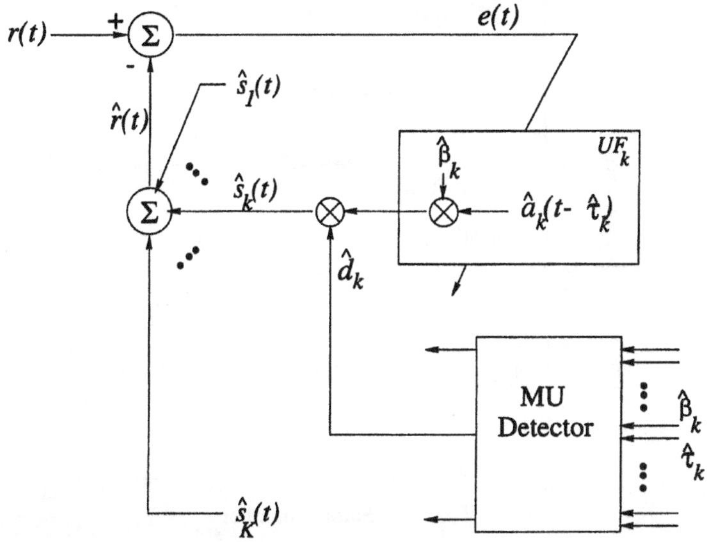

Figure 6.9. **Operation of multiuser parameter estimation receiver. Data bits are available from decision-directed output.**

value for convergence. However, many multiuser parameter estimators are found in the literature, such as those discussed at the beginning of this section. These methods can usually provide a delay estimate near the true delay and good channel amplitude estimates. For the simulation results, we assume such an initial estimator is used to start the tracking algorithm fairly close to the true values. Furthermore, we note that the data bits, $d_{k,m}(l)$, are not included in the estimation process, but are assumed unknown *a priori*. In the simulations, we assume that the data bits are available from decision-directed adaptation, where the symbols $d_{k,m}(l)$ are replaced by the decisions $\hat{d}_{k,m}(l)$ (Fig. 6.9).

Figs. 6.10 and 6.11 show the estimation error for constant amplitudes and time delays for perfectly power-controlled (equal power) users. It is seen that both users are able to accurately converge to the correct amplitudes and delays. The results for the same set of assumptions, but with a near-far ratio of 20 dB are shown in Figs. 6.12 and 6.13. Again, the estimator is able to accurately converge to the correct values of the parameters. The ability of the estimator to track time-varying parameters is shown in Fig. 6.14 where the time-delays linearly decrease (increase) for the weaker (stronger) user. The near-far ratio was 20 dB. Although the amount of change of the time delays for each user does not

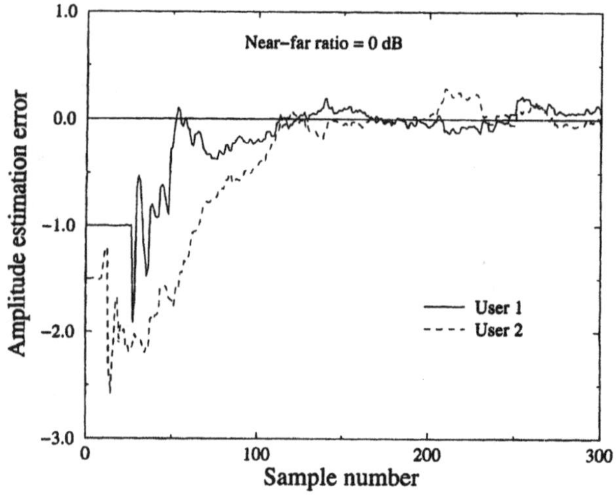

Figure 6.10. Parameter estimation errors for constant channel amplitudes with a near-far ratio of 0 dB. The SNR for each user is 10 dB.

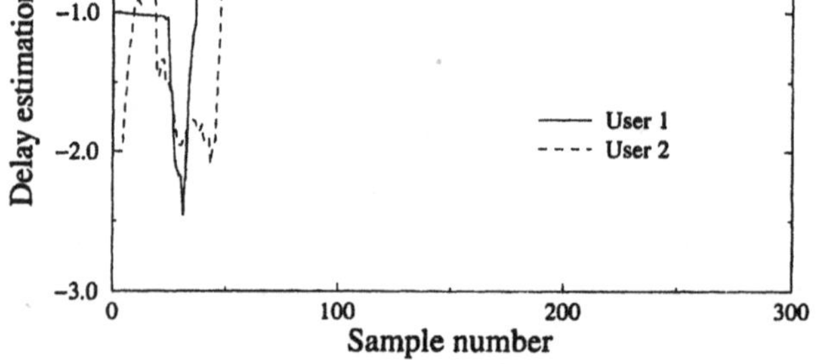

Figure 6.11. Parameter estimation errors for constant time delays with a near-far ratio of 0 dB. The SNR for each user is 10 dB.

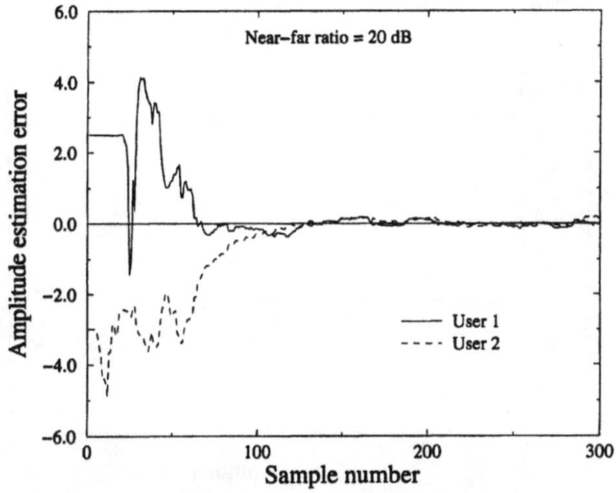

Figure 6.12. **Parameter estimation errors for constant channel amplitudes with a near-far ratio of 20 dB. User 1 is the weaker user with a SNR of 10 dB.**

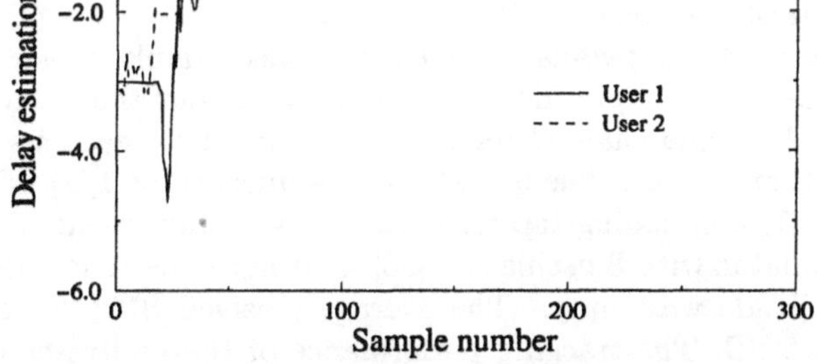

Figure 6.13. **Parameter estimation errors for constant time delays with a near-far ratio of 20 dB. User 1 is the weaker user with a SNR of 10 dB.**

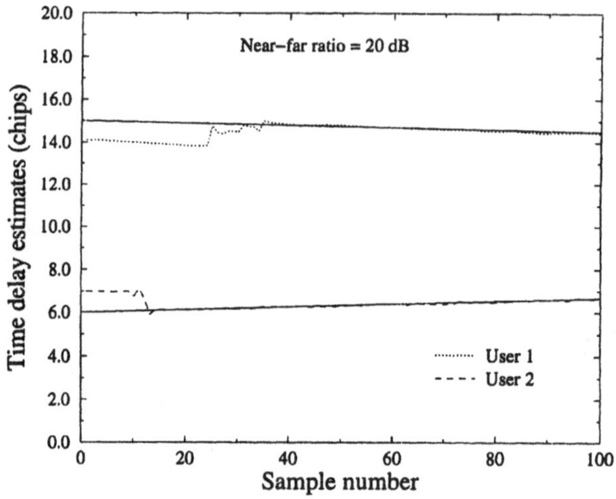

Figure 6.14. The tracking ability of the estimator for time-varying delays at a near-far ratio of 20 dB. User 1 is the weaker user with a SNR of 10 dB.

appear to be significant in the figure, we note that the users' time delays change by $0.5T_c$ and $0.67T_c$ over 100 samples, with $q = 2$. Assuming the propagation speed is the speed of light in a vacuum (3×10^8 m/s) and that $T_c = 0.001$ sec, then each user has moved distances of 150,000 m and 200,000 m over a time span of 50 msec. Such great changes of distance in such a short time span are not likely to occur in practical situations. However, the estimator was able to converge and track the parameters in such a harsh scenario.

The tracker for a two-user system was also simulated for a fading channel where the amplitudes are time-varying, but the delays remain constant. A simple channel model was assumed for each user with a single tap (flat fading). The bit rate was assumed to be $1/T_b = 1000$ bps with $N = 31$. The fading tap for each user was implemented by Jakes' fading simulator with 8 oscillators [80], and normalized so that the average amplitude was unity. The average received SNR for each user was set to 5 dB. The tracking performance of the multiuser tracker is shown in Fig. 6.15 for each user with a Doppler frequency of 50 Hz for User 1 and 100 Hz for User 2. The figure shows the magnitude of each user's complex channel amplitudes. As the figure indicates, the estimator/tracker is able to accurately track the time-varying ampli-

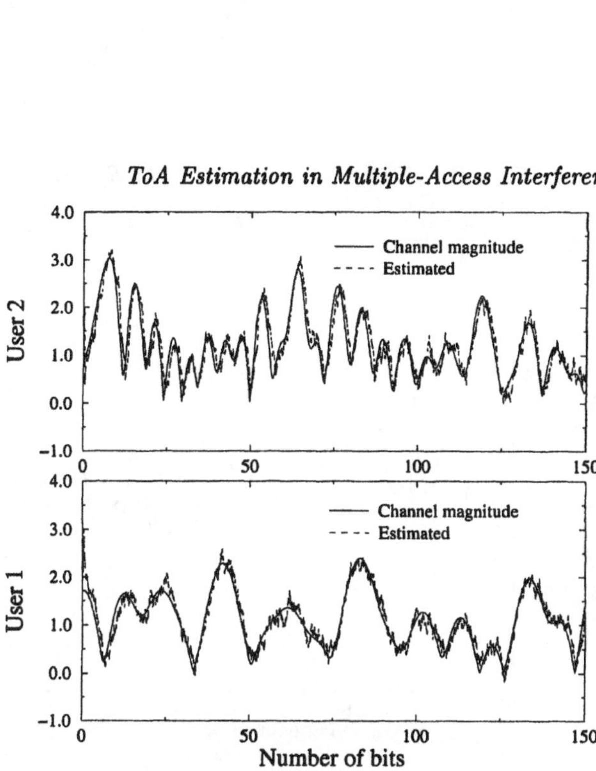

Figure 6.15. Time-varying channel amplitude tracking using the multiuser parameter tracker.

tudes of each user, even for fast fading rates of 100 Hz (Doppler). We do note, however, that the variance of the estimator does appear to increase slightly as the fade rate increases. Also, although the average near-far ratio is 0 dB, the instantaneous ratio varies drastically due to the fading of the channel since there is no power control. As shown in Fig. 6.16, the instantaneous near-far ratio, with respect to the second user, varies at any time from -20 dB to +20 dB. In spite of the varying powers, the tracker is still able to provide excellent performance.

To further quantify the performance of the estimator, the mean-squared-error (MSE) from simulation of the estimator is compared to that predicted by the CRLB. The CRLB was evaluated for a two user system where user-1 was the desired user. In near-far situations, the first user is the weaker user. For comparison with the CRLB results, an ensemble average squared error was computed for the estimates as follows:

$$\mathrm{MSE}_{\beta}(n) = \frac{1}{N_s}\sum_{i=1}^{N_s}\left[\beta_1 - \hat{\beta}_1^{(i)}(n)\right]^2 \qquad (6.90)$$

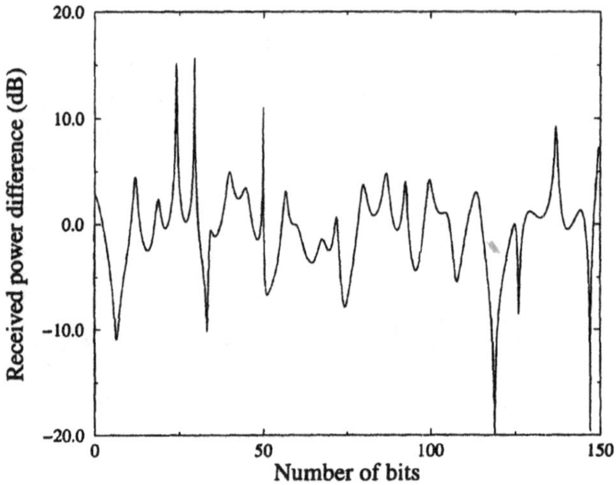

Figure 6.16. Instantaneous near-far ratio for the fading channel, with respect to the second user, and with no power control.

$$\mathrm{MSE}_{\tau}(n) = \frac{1}{N_s}\sum_{i=1}^{N_s}\left[\tau_1 - \hat{\tau}_1^{(i)}(n)\right]^2 \tag{6.91}$$

where $\mathrm{MSE}_{\beta}(n)$ and $\mathrm{MSE}_{\tau}(n)$ denote the MSE of the amplitude and delay, respectively, of the desired user (user-1) at iteration n of the estimator. N_s is the number of ensemble samples used to form the MSE and $\hat{\beta}_1^{(i)}(n)$ and $\hat{\tau}_1^{(i)}(n)$ are the estimates of the amplitude and delay at time n of the ith member in the ensemble. Note that the CRLB is dependent on the realizations of d_{k,m_l} and $\beta_k(l)$. Thus, the CRLB that is plotted in the remaining figures is averaged over several simulation runs.

The performance of the estimator compared to the CRLB is first evaluated versus the SNR of user-1. The number of ensemble samples was chosen to be $N_s = 300$. The MSE terms defined in (6.90) and (6.91) were computed for the parameter estimates formed after 70 iterations of the filter. Figs. 6.17 and 6.18 show the MSE of the estimates of the amplitudes and time delays, respectively, against the SNR of the first user. Also shown is the CRLB for the first user. Curves are shown for near-far ratios of 0 dB and 20 dB. We find that the computed MSE is

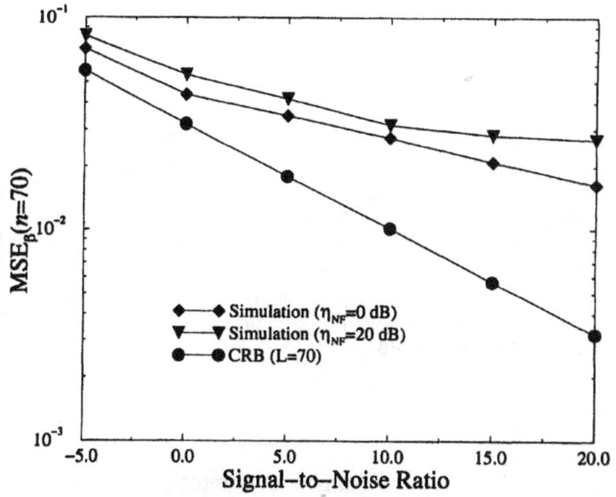

Figure 6.17. The MSE of the amplitude estimates versus SNR from the simulated estimator and the CRLB for different near-far ratios.

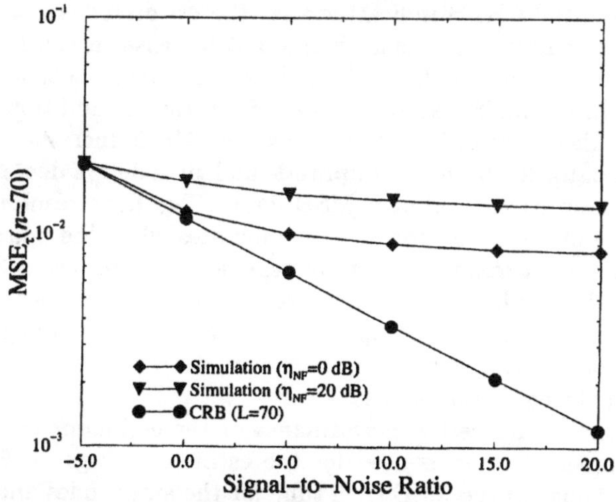

Figure 6.18. The MSE of the time delay estimates versus SNR from the simulated estimator and the CRLB for different near-far ratios.

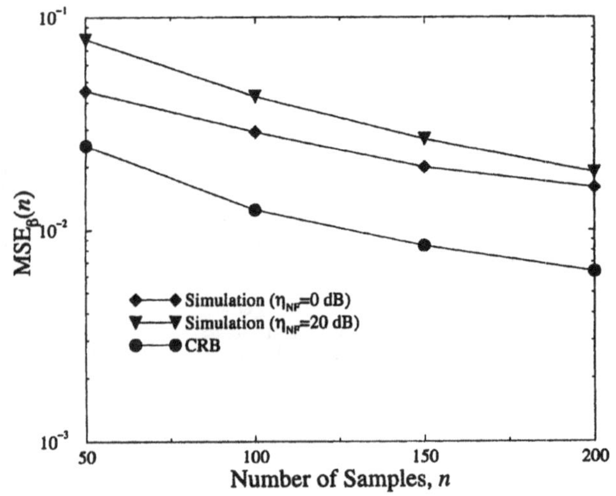

Figure 6.19. The MSE of the amplitude estimates versus the number samples from the simulated estimator and the CRLB for different near-far ratios.

greater than the CRLB indicating that the estimator is not *efficient*[1]. The figures indicate that there is a slight increase in the MSE as the near-far ratio is increased. We would expect a truly near-far resistant estimator to be unaffected by the near-far ratio. In addition, although the CRLB decreases quickly to zero as the SNR is increased, the MSE of the estimator for both the amplitude and time delays decreases much slower and seems to exhibit a MSE floor. The floor results from the interference of the other user which dominates when the noise power is low. Since the estimator is not fully near-far resistant, the MSE floor is produced. It is evident that the increased power of the second user is treated as increased noise compared to the case of equal power users. If the MSE for the stronger user were shown, we would find a decrease in MSE from the equal power case.

Finally, we compare the performance of the estimator as a function of the number of samples used for the estimates. Figs. 6.19 and 6.20 show the change of the MSE with time for the amplitudes and delays of the weaker user, respectively. Again, the number of ensemble samples

[1] An efficient estimator is one that satisfies the CRLB with equality [5].

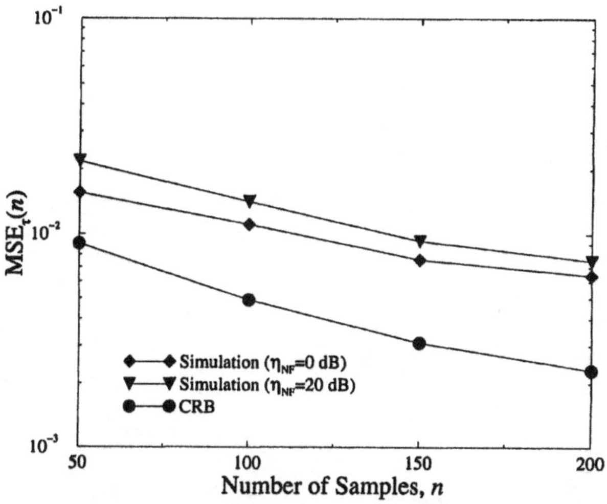

Figure 6.20. The MSE of the time delay estimates versus the number of samples from the simulated estimator and the CRLB for different near-far ratios.

used to form the MSE estimates was 300. The SNR of the weaker user is 5 dB. As expected, the MSE is reduced as the number of samples is increased. However, the MSE for the estimator will approach the steady state for the MSE, while the CRLB continues to zero. Also, as the filter is allowed to track the channel amplitudes and delays, the MSE for the near-far scenario ($\eta_{NF} = 20$ dB) begins to approach the MSE for the equal power scenario. Thus, the figures indicate that allowing the filter to operate for a long enough period of time results in an MSE that is independent of the powers of the users.

3. SUMMARY

This chapter has evaluated the effect that MAI has on conventional methods of time delay estimation. The analysis showed that MAI has a drastic affect on the performance of the conventional NC-DLL. If these methods are used to provide ToA estimates for location, the accuracy of the location method will suffer also. A few estimators were reviewed which sought to accurately estimate and/or track the delay in MAI. A method for tracking time-varying delays and channel amplitudes was developed and simulated to show good performance in MAI.

Chapter 7

ANALYSIS AND MITIGATION OF NLOS EFFECTS

NLoS propagation has been identified as one of the primary factors that limit the accuracy of the time-based location systems [16, 18, 19, 154, 197]. The typical error introduced by NLoS propagation has been measured in the GSM system which indicates that the NLoS error can average between 500-700 meters [154]. These results also apply to CDMA and other cellular systems as well since they are dependent on the physical environment, not the system deployed. For time-based systems, NLoS propagation introduces a bias to the ToA and TDoA measurements which the location algorithms are not designed to handle. Consequently, the location estimates produced by NLoS corrupted measurements will be in error from the true position of the MS.

All of the radiolocation techniques discussed in Chapter 3 require that a LoS path exist between the MS and BSs in order to obtain high accuracy. However, in cellular networks, it is unlikely that LoS path exists between a MS and its serving BS, especially in macrocells, not withstanding the surrounding BSs that are needed for location. Consequently, the measured ToAs (or TDoAs) reflect propagation distances that are greater than the actual distances between the MS and each BS, as was illustrated in Fig. 3.8.

Several authors have presented algorithms for computing location estimates from noisy TDoA or ToA measurements based on linear and nonlinear LS methods as reviewed in Chapter 4. For accurate location estimates, these algorithms usually assume that a direct path exists between the transmitter and receiver. The LS methods are used to calculate the position of a MS from a single measurement at several BSs. Multiple measurements can be included by increasing the length of the measurement vector. In order to track the position, iterative procedures such as recursive least squares (RLS) or EKF can be employed.

In this chapter, the effect of the bias that is introduced into the time-based location algorithms by NLoS propagation is analyzed. Two of the more popular algorithms in the literature, LS with Taylor series linearization (ToA and TDoA) and the pseudomeasurement algorithm (TDoA only), are studied in the analysis. In Section 1, the bias that results in the location estimate due to NLoS propagation is mathematically derived for each algorithm. In Section 2, methods to mitigate the NLoS problem are examined.

1. ANALYSIS OF NLOS EFFECTS

NLoS propagation corrupts the ToA and TDoA measurements that are made at the BSs. Determining the extent to which accuracy is affected is necessary in order to understand whether these location methods can be employed "as is" in wireless radio systems. If they are robust, then little needs to be changed. However, if the errors are significant, then new approaches need to be developed to mitigate the NLoS effects.

1.1 TIME-BASED RADIOLOCATION METHODS

A brief review of the LS algorithms based on linearization and pseudomeasurements is necessary before beginning the analysis. Below, we provide more details of the two algorithms than were given in Chapter 4.

1.1.1 TOA AND TDOA EQUATIONS

The ToA location method considered in this chapter is based on the estimation of the time of transmission, t_s, as well as the location of the MS, $\boldsymbol{x}_s$. For TDoA, the time of transmission need not be known, so only $\boldsymbol{x}_s$ is estimated. The generic parameter vector $\boldsymbol{\theta}_s$ will be used to denote the unknown quantities for both location schemes in order to present general results. However, it should be kept in mind that $\boldsymbol{\theta}_s = [\boldsymbol{x}_s^T, t_s]^T$ for ToA and $\boldsymbol{\theta}_s = \boldsymbol{x}_s$ for TDoA.

Recall from Chapter 4 that the arrival time of a signal is modeled by

$$t_i(\boldsymbol{\theta}_s) = t_s + D_i(\boldsymbol{x}_s)/c \qquad i = 1, \ldots, N_B \tag{7.1}$$

where (7.1) differs from (4.39) by the inclusion of the unknown time of transmission. The TDoAs are modeled by

$$\rho_{i,1}(\boldsymbol{\theta}_s) = \frac{D_i(\boldsymbol{x}_s) - D_1(\boldsymbol{x}_s)}{c} \qquad i = 2, \ldots, N_B \tag{7.2}$$

where $D_i(\boldsymbol{x}_s)$ is the distance between the MS and i^{th} BS and, without loss of generality, we have referenced the TDoAs to the first BS.

1.1.2 LOCATION ESTIMATION ALGORITHMS

The methods to be examined use the LS technique to form the location estimate from a single set of noisy measurements from N_B BSs [43, 44, 158, 180]. Recall from equation (4.62) that the $N_B \times 1$ vector of noisy measurements, in general form, can be expressed as

$$\boldsymbol{r}_m = \boldsymbol{C}(\boldsymbol{\theta}_s) + \boldsymbol{n}_m \tag{7.3}$$

where $\boldsymbol{r}_m$ is the vector of measurements, either ToA or TDoA. Note that $\boldsymbol{r}_m$ is dependent on the unknown parameter vector $\boldsymbol{\theta}_s$, although it is not shown explicitly above. The function $\boldsymbol{C}(\boldsymbol{\theta}_s)$ depends on whether the location algorithm uses ToA or TDoA measurements. For ToA

$$\boldsymbol{C}(\boldsymbol{\theta}_s) = t_s\boldsymbol{1} + \boldsymbol{D}(\boldsymbol{\theta}_s)/c \tag{7.4}$$

according to (4.67), and for TDoA

$$\boldsymbol{C}(\boldsymbol{\theta}_s) = \boldsymbol{\mathcal{R}}(\boldsymbol{\theta}_s) \tag{7.5}$$

according to (4.63). The equations for $\boldsymbol{D}(\boldsymbol{\theta}_s)$ and $\boldsymbol{\mathcal{R}}(\boldsymbol{\theta}_s)$ are given in (4.64) and (4.65), respectively, and are repeated here for the reader's convenience:

$$\boldsymbol{D}(\boldsymbol{\theta}_s) = \begin{bmatrix} D_1 \\ D_2 \\ \vdots \\ D_{N_B} \end{bmatrix} \quad \text{and} \quad \boldsymbol{\mathcal{R}}(\boldsymbol{\theta}_s) = \begin{bmatrix} \rho_{2,1} \\ \rho_{3,1} \\ \vdots \\ \rho_{N_B,1} \end{bmatrix} \tag{7.6}$$

where the dependence of D_i and $\rho_{i,1}$ on $\boldsymbol{\theta}_s$ has been removed to simplify the notation.

The weighted least squares (WLS) estimate is then formed by minimizing the quadratic cost function defined along the lines of equation (5.6):

$$\mathcal{E}(\boldsymbol{\theta}_s) = [\boldsymbol{r}_m - \boldsymbol{C}(\boldsymbol{\theta}_s)]^T \boldsymbol{W} [\boldsymbol{r}_m - \boldsymbol{C}(\boldsymbol{\theta}_s)] \ . \tag{7.7}$$

The weight matrix $\boldsymbol{W}$ can be chosen to achieve the ML estimate with $\boldsymbol{W} = \boldsymbol{\Sigma}_m^{-1}$. In the analysis that follows, we will assume that $\boldsymbol{W} = \boldsymbol{I}$ for simplicity. We will discuss using the weight matrix to combat NLoS error in Section 2.

We note that $\boldsymbol{C}(\boldsymbol{\theta}_s)$ is a nonlinear vector function in the unknown $\boldsymbol{x}_s$. In Chapter 5, a nonlinear LS method to solve for the unknown quantity was applied. A gradient search minimization technique was used to find the minimum of $\mathcal{E}(\boldsymbol{\theta}_s)$. Another approach that was mentioned in Chapter 4 linearizes $\boldsymbol{C}(\boldsymbol{\theta}_s)$ about a reference point [43, 180]. We will examine the this linearization approach for ToA and TDoA because it is a popular algorithm and because of its simplicity which makes the following analysis of NLoS bias effects more tractable.

In Chapter 4, it was shown that using a Taylor series expansion of $\boldsymbol{C}(\boldsymbol{\theta}_s)$ about some $\boldsymbol{\theta}_o$ and keeping only the first two terms, the LS estimator can be shown to be equal to [180]

$$\hat{\boldsymbol{\theta}}_{\text{TS}} = \boldsymbol{\theta}_o + [\boldsymbol{H}_C^T \boldsymbol{H}_C]^{-1} \boldsymbol{H}_C^T [\boldsymbol{r}_m - \boldsymbol{C}(\boldsymbol{\theta}_o)] \tag{7.8}$$

where $\boldsymbol{H}_C$ is the $N_B \times n$ Jacobian matrix of $\boldsymbol{C}(\boldsymbol{\theta}_s)$ evaluated at $\boldsymbol{\theta}_o$, with $n = 3$ for ToA and $n = 2$ for TDoA. The Jacobian matrix can be shown to be

$$\begin{aligned} \boldsymbol{H}_C &= \left.\frac{\partial \boldsymbol{C}(\boldsymbol{\theta}_s)}{\partial \boldsymbol{\theta}_s}\right|_{\boldsymbol{\theta}_s=\boldsymbol{\theta}_o} = \left[\begin{array}{ccc} \frac{\partial \boldsymbol{C}(\boldsymbol{\theta}_s)}{\partial x_s} & \frac{\partial \boldsymbol{C}(\boldsymbol{\theta}_s)}{\partial y_s} & \frac{\partial \boldsymbol{C}(\boldsymbol{\theta}_s)}{\partial t_s} \end{array}\right]\Bigg|_{\boldsymbol{\theta}_s=\boldsymbol{\theta}_o} \\ &= \frac{1}{c}\left[\begin{array}{cc} \frac{1}{D_1(\boldsymbol{\theta}_o)}[\boldsymbol{x}_o - \boldsymbol{x}_1]^T & c \\ \vdots & \vdots \\ \frac{1}{D_{N_B}(\boldsymbol{\theta}_o)}[\boldsymbol{x}_o - \boldsymbol{x}_{N_B}]^T & c \end{array}\right] \end{aligned} \tag{7.9}$$

for ToA, and for TDoA

$$\boldsymbol{H}_C = \frac{1}{c}\left[\begin{array}{c} \frac{1}{D_2(\boldsymbol{\theta}_o)}[\boldsymbol{\theta}_o - \boldsymbol{x}_2]^T \\ \vdots \\ \frac{1}{D_{N_B}(\boldsymbol{\theta}_o)}[\boldsymbol{\theta}_o - \boldsymbol{x}_{N_B}]^T \end{array}\right] - \frac{1}{cD_1(\boldsymbol{\theta}_o)}\,\mathbf{1}\,[\boldsymbol{\theta}_o - \boldsymbol{x}_1]^T . \tag{7.10}$$

If we cannot be confident in the linear approximation by Taylor series expansion, an iterative approach can be used [43] for $\boldsymbol{\theta}_s$; however, we will assume that $\boldsymbol{\theta}_o$ is close enough to $\boldsymbol{\theta}_s$ that the approximation is accurate.

In order to avoid errors with linearization in the location algorithm as applied to TDoA, alternative algorithms have been developed [44, 136]. As discussed in Chapter 4, these algorithms estimate the location through the use of pseudomeasurements, which are actually known functions of the TDoA measurements. The pseudomeasurements are formed by squaring the TDoA measurements and algebraically manipulating the equations to arrive at the result in (4.79) which is repeated here:

$$\boldsymbol{\varphi} = \boldsymbol{\Delta}\boldsymbol{\theta}_s + D_1(\boldsymbol{\theta}_s)\boldsymbol{r}_m \tag{7.11}$$

where the vector $\boldsymbol{\varphi}$ and the matrix $\boldsymbol{\Delta}$ are defined in equations (4.77) and (4.76), respectively. The $N_B - 1$ vector $\boldsymbol{\varphi}$ contains the pseudomeasurements.

Following the procedure in Chapter 4, the LS solution obtained from the pseudomeasurement algorithm is

$$\hat{\theta}_{\text{PM}} = \hat{\boldsymbol{x}}_s = \left[\boldsymbol{\Delta}^T \boldsymbol{P}^T \boldsymbol{P} \boldsymbol{\Delta}\right]^{-1} \boldsymbol{\Delta}^T \boldsymbol{P}^T \boldsymbol{P} \boldsymbol{\varphi} \ . \tag{7.12}$$

where $\boldsymbol{P} = (\boldsymbol{I} - \boldsymbol{Z})[\text{diag}(\boldsymbol{r}_m)]^{-1}$ is a projection matrix with $\boldsymbol{r}_m$ in its null space and $\boldsymbol{Z}$ is a circular shift matrix.

It is interesting to note that the methods presented above are appropriate when a single measurement is available at each BS. More measurements could be incorporated by increasing the length of the vector $\boldsymbol{D}$ for ToA and the vector $\boldsymbol{\mathcal{R}}$ or $\boldsymbol{\varphi}$ for TDoA and making the necessary algorithmic adjustments. When it is desired to track the position of a moving MS, frequent iterations of the algorithm are necessary. More appropriate methods for tracking are based on iterative algorithms such as the RLS algorithm or Kalman filtering. Due to some fundamental similarities between the RLS and the Kalman filtering algorithms, we will limit this brief discussion to the linear KF and the EKF.

Application of the Kalman filtering algorithms is a simple extension to the above derivations. To apply the Kalman filtering algorithms, we must first specify the state and measurement models. The state model can assume the form

$$\boldsymbol{\theta}_s(k+1) = \boldsymbol{\Phi}_s \boldsymbol{\theta}_s(k) + \boldsymbol{w}_s(k) \tag{7.13}$$

where $\boldsymbol{w}_s(k)$ is a zero-mean Gaussian noise process with covariance $\boldsymbol{Q}_s = \sigma_w^2 \boldsymbol{I}$. The measurement models have already been given in (7.1) for ToA and in (7.2) or (7.11) for TDoA depending on which algorithm is used. With these models, straightforward application of the standard EKF or linear KF are all that is required. The analysis that follows will not address KF or RLS implementations for location tracking. The details of the KF filter equations are summarized in Appendix B.

1.2 EFFECT OF NLOS ERROR

1.2.1 DERIVATIONS

The error from NLoS propagation manifests itself as a bias on the ToA and TDoA measurements. Hence, the measurement model presented in equation (7.3) must be altered according to the following:

$$\boldsymbol{r}_m = \boldsymbol{C}(\boldsymbol{\theta}_s) + \boldsymbol{b} + \boldsymbol{n}_m \tag{7.14}$$

where the bias term $\boldsymbol{b}$ has been added and the appropriate replacements are made for $\boldsymbol{r}_m$ and $\boldsymbol{C}(\boldsymbol{\theta}_s)$ for ToA and TDoA. It has been suggested that the TDoA method provides some inherent protection from NLoS bias. This is because of the differencing that is used. A better understanding is obtained if we view TDoAs as simply the differences of ToAs. If the ith BS and the reference BS experience bias of similar magnitude, differencing will nearly eliminate the bias seen by the TDoA algorithm.

To compute the effect of the NLoS error on the first algorithm, we use the standard definition of the estimator bias given by $\mathcal{B} = \mathrm{E}\left\{\hat{\boldsymbol{\theta}}\right\} - \boldsymbol{\theta}$ [167]. Applying this definition to the estimator of equation (7.8), we get

$$\begin{aligned}\mathcal{B}_{\mathrm{TS}} &= \mathrm{E}\left\{\hat{\boldsymbol{\theta}}_{\mathrm{TS}}\right\} - \boldsymbol{\theta}_s \\ &= \left[\boldsymbol{H}_C^T\boldsymbol{H}_C\right]^{-1}\boldsymbol{H}_C^T[\boldsymbol{C}(\boldsymbol{\theta}_s) - \boldsymbol{C}(\boldsymbol{\theta}_o) - \boldsymbol{H}_C(\boldsymbol{\theta}_s - \boldsymbol{\theta}_o) \\ &\qquad + \mathrm{E}\{\boldsymbol{b}\}] \qquad (7.15)\end{aligned}$$

where we have used $\mathrm{E}\{\boldsymbol{n}_m\} = \mathbf{0}$. The first three terms in the brackets are the error introduced by linearization which we will assume to be negligible so that the bias is given by

$$\mathcal{B}_{\mathrm{TS}} = \left[\boldsymbol{H}_C^T\boldsymbol{H}_C\right]^{-1}\boldsymbol{H}_C^T\mathrm{E}\{\boldsymbol{b}\} \ . \qquad (7.16)$$

Thus, if the measurement bias is small, the resulting location estimate bias will also be small. Note also that the bias is dependent on the choice of $\boldsymbol{\theta}_o$, the point around which the Jacobian matrix is formed, through the quantity $\boldsymbol{H}_C$.

The computation of the bias introduced by NLoS error into the pseudomeasurement algorithm, based on TDoA, is more complicated since the measured TDoAs, which are corrupted by the bias, are squared to form the pseudomeasurements, $\boldsymbol{\varphi}$. In the following, the bias caused by the zero-mean measurement noise, $\boldsymbol{n}_m$ is assumed to be negligible compared to the NLoS bias, $\boldsymbol{b}$. This is a reasonable assumption as the bias associated with the ith BS, b_i, was shown in [154] to be much greater than the measurement noise, $\boldsymbol{n}_m$.

Let the measured TDoA be represented as $\hat{\boldsymbol{\rho}} = \boldsymbol{\rho} + \boldsymbol{b}$, where $\boldsymbol{\rho}$ is the true value. The location estimate when $\hat{\boldsymbol{\rho}}$ is used instead of $\boldsymbol{\rho}$ is given by $\hat{\boldsymbol{\theta}}_{\mathrm{PM}} = \boldsymbol{\theta}_s + \tilde{\boldsymbol{\theta}}_{\mathrm{PM}}$, where $\tilde{\boldsymbol{\theta}}_{\mathrm{PM}}$ is the error. Likewise, the biased range estimate is denoted by $\hat{D}_1 = D_1 + \tilde{D}_1$. Thus, we seek to compute the bias $\mathcal{B}_{\mathrm{PM}} = \mathrm{E}\{\tilde{\boldsymbol{\theta}}_{\mathrm{PM}}\} = \tilde{\boldsymbol{\theta}}_{\mathrm{PM}}$ since we are ignoring random measurement noise, so that $\boldsymbol{b}$ is assumed constant. Inserting $\hat{\boldsymbol{\theta}}_{\mathrm{PM}}$ into equation (7.11), we have

$$\begin{aligned}\boldsymbol{\Delta}\hat{\boldsymbol{\theta}}_{\mathrm{PM}} &= \hat{\boldsymbol{\varphi}} - \hat{D}_1\hat{\boldsymbol{\rho}} \qquad (7.17)\\ &= [\boldsymbol{m} - D_1\boldsymbol{\rho}] - \left[\left(\frac{1}{2}\boldsymbol{b} + \boldsymbol{\rho} + \hat{D}_1\mathbf{1}\right) \odot \boldsymbol{b} + \tilde{D}_1\boldsymbol{\rho}\right] \qquad (7.18)\end{aligned}$$

where $\boldsymbol{u} \odot \boldsymbol{v}$ denotes the component by component product of the vectors $\boldsymbol{u}$ and $\boldsymbol{v}$. Since $\Delta\hat{\boldsymbol{\theta}}_{\text{PM}} = \Delta\boldsymbol{\theta}_s + \Delta\tilde{\boldsymbol{\theta}}_{\text{PM}}$, we see that

$$\Delta\tilde{\boldsymbol{\theta}}_{\text{PM}} = -\left[\left(\frac{1}{2}\boldsymbol{b} + \boldsymbol{\rho} + \hat{D}_1\mathbf{1}\right) \odot \boldsymbol{b} + \tilde{D}_1\boldsymbol{\rho}\right] . \tag{7.19}$$

Projecting the matrix equation above with the projection $\boldsymbol{P}$ gives

$$\boldsymbol{P}\Delta\tilde{\boldsymbol{\theta}}_{\text{PM}} = -\boldsymbol{P}\left[\left(\frac{1}{2}\boldsymbol{b} + \boldsymbol{\rho} + \hat{D}_1\mathbf{1}\right) \odot \boldsymbol{b}\right] . \tag{7.20}$$

Thus, to determine the bias $\mathcal{B}_{\text{PM}} = \tilde{\boldsymbol{\theta}}_{\text{PM}}$, we must determine $\hat{D}_1$. An expression for $\hat{D}_1$ was derived in [158] where it was shown that $D_1 = \boldsymbol{\rho}^T\boldsymbol{P}_\Delta\boldsymbol{\rho}/\boldsymbol{\rho}^T\boldsymbol{P}_\Delta\boldsymbol{\rho}$ where $\boldsymbol{P}_\Delta = \boldsymbol{I} - \boldsymbol{\Delta}[\boldsymbol{\Delta}^T\boldsymbol{\Delta}]^{-1}\boldsymbol{\Delta}^T$. Using the measured values of $\boldsymbol{\rho}$ and $\boldsymbol{\varphi}$, we can show that

$$\begin{aligned} \hat{D}_1 &= \frac{\hat{\boldsymbol{\rho}}^T\boldsymbol{P}_\Delta\hat{\boldsymbol{\varphi}}}{\hat{\boldsymbol{\rho}}^T\boldsymbol{P}_\Delta\hat{\boldsymbol{\rho}}} \\ &= \frac{[\boldsymbol{\rho}+\boldsymbol{b}]^T\boldsymbol{P}_\Delta\left[\boldsymbol{\varphi} - \left(\boldsymbol{\rho} + \frac{1}{2}\boldsymbol{b}\right)\odot\boldsymbol{b}\right]}{\boldsymbol{\rho}^T\boldsymbol{P}_\Delta\boldsymbol{\rho} + 2\boldsymbol{\rho}^T\boldsymbol{P}_\Delta\boldsymbol{b} + \boldsymbol{b}^T\boldsymbol{P}_\Delta\boldsymbol{b}} . \end{aligned} \tag{7.21}$$

With this result, we can compute the bias, $\mathcal{B}_{\text{PM}} = \tilde{\boldsymbol{\theta}}_s$, from equation (7.19) which gives

$$\mathcal{B}_{\text{PM}} = -\left[\boldsymbol{\Delta}^T\boldsymbol{P}^T\boldsymbol{P}\boldsymbol{\Delta}\right]^{-1}\boldsymbol{\Delta}^T\boldsymbol{P}^T\boldsymbol{P}\left[\left(\frac{1}{2}\boldsymbol{b} + \boldsymbol{\rho} + \hat{D}_1\mathbf{1}\right) \odot \boldsymbol{b}\right] \tag{7.22}$$

which requires substitution of $\hat{D}_1$ from (7.21).

1.2.2 DISCUSSION AND COMPARISON

In order to see quantitatively the effect of the bias introduced by NLoS propagation, we consider a BS layout according to Fig. 7.1, where we have shown five BSs and the position of the MS. It has been shown experimentally that the NLoS error on range estimates is randomly distributed with support on the interval [0 m, 1300 m] with a mean of approximately 500 m in GSM systems [154]. We would expect that this would be the case in other systems (e.g., CDMA and TDMA) as well since the NLoS error is a function of the environment, and not of the system implementation, as the standard measurement noise would be. Table 7.1 shows an example comparison of Taylor series algorithms for both ToA and TDoA (labeled $\mathcal{B}_{\text{TS}}$) and the pseudomeasurement algorithm (labeled $\mathcal{B}_{\text{PM}}$) for different numbers of BSs and NLoS error vectors. The results for the LS algorithm based on Taylor series expansion only used a single iteration.

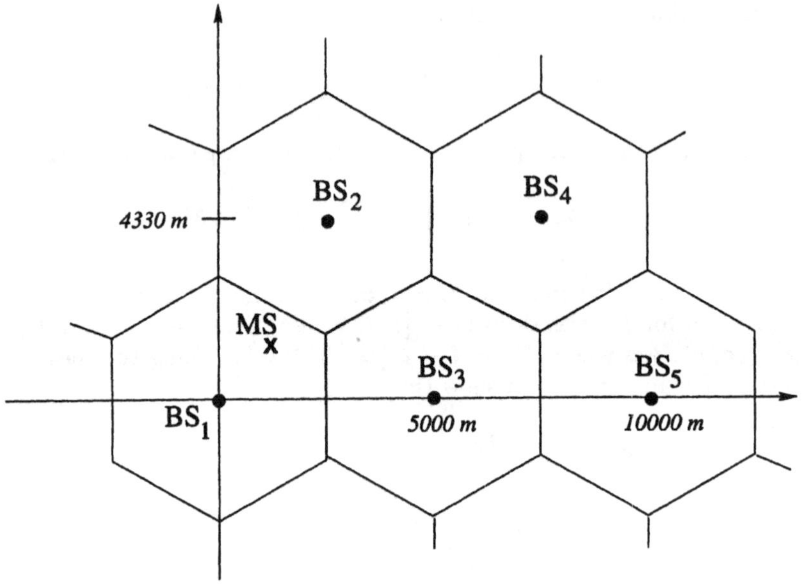

Figure 7.1. Typical macrocellular BS deployment scenario with a hexagonal cell radius of 2500 m.

Table 7.1. Comparison of the location estimate bias for ToA and TDoA algorithms based on Taylor series expansion, and the TDoA algorithm based on pseudomeasurements.

	ToA	*TDoA*	
	$\mathcal{B}_{TS}$	$\mathcal{B}_{TS}$	$\mathcal{B}_{PM}$
$N_B = 3$	$\begin{bmatrix} -151 \\ -263 \\ 659/c \end{bmatrix}$	$\begin{bmatrix} -150 \\ -262 \end{bmatrix}$	N/A
$N_B = 4$	$\begin{bmatrix} 63 \\ -172 \\ 1183/c \end{bmatrix}$	$\begin{bmatrix} -372 \\ -423 \end{bmatrix}$	$\begin{bmatrix} -873 \\ -465 \end{bmatrix}$
$N_B = 5$	$\begin{bmatrix} 373 \\ 454 \\ 1737/c \end{bmatrix}$	$\begin{bmatrix} -695 \\ 62 \end{bmatrix}$	$\begin{bmatrix} 205 \\ 65 \end{bmatrix}$

The TDoAs, for simplicity, are computed from the differences of measured ToAs relative to the first BS. A NLoS error vector, $\boldsymbol{b}$, is added to each ToA; thus, the NLoS error vector for the TDoA algorithms can be formed by differencing the elements in $\boldsymbol{b}$. The system measurement noise is assumed to be zero in order for the location error to be completely determined by the NLoS bias. The NLoS error vectors for $N_B = 3, 4, 5$ are:

$$N_B = 3: \quad \boldsymbol{b} = [350, 600, 500]^T$$
$$N_B = 4: \quad \boldsymbol{b} = [350, 600, 500, 450]^T$$
$$N_B = 5: \quad \boldsymbol{b} = [350, 600, 500, 450, 700]^T$$

For each N_B, the MS was located at $\boldsymbol{x}_s = [1000, 1000]^T$, and $t_s = 50/c$ for ToA. For the Taylor series algorithms, the initial guess, $\boldsymbol{x}_o$, for the location was taken as the center of the N_B BSs used for location and $t_o = 48/c$. The table shows that the Taylor series based algorithm experiences much less of a bias than the pseudomeasurement algorithm. This is because of the squaring of TDoA measurements that is done to form the pseudomeasurements. Also, the TDoA-based location biases appear to be worse than those produced by the ToA algorithm.

To further illustrate the effect of the NLoS bias on location estimation error, Figs. 7.2 and 7.3 plot the cumulative distribution functions (*cdf*'s) for 10000 simulated location and bias vector scenarios for both the Taylor series and the pseudomeasurement algorithms. The location of the MS was randomly placed among the BSs in Fig. 7.1. The 3, 4, or 5 BSs nearest to the MS were used for the location process. Since no measurement noise is considered, the location error is simply given by the location estimate bias. The NLoS bias for each BS was chosen to be randomly distributed according to a clipped Gaussian distribution with support on [0 m, 1300 m] and a mean and standard deviation of 500 m and 436 m, respectively, as indicated by the measurements in [154]. Fig. 7.2 shows that the ToA and TDoA algorithms based on the Taylor series approximation perform nearly the same for 3 BSs; however, for more than 3 BSs, the ToA algorithm performs slightly better than the TDoA algorithm, on average, in the presence of NLoS propagation error. The mean location error, as indicated by the curves, ranges from 250 m for ToA with 5 BSs to 400 m for ToA with 3 BSs. Thus, more information through the use of more BSs, even though their measurements may be greatly in error, improves the performance. The FCC requirements of 125 m accuracy are only met between 12% to 19% of the time depending on the number of BSs used.

Fig. 7.3 indicates that the pseudomeasurement algorithm performs much worse with NLoS corrupted measurements than the Taylor series

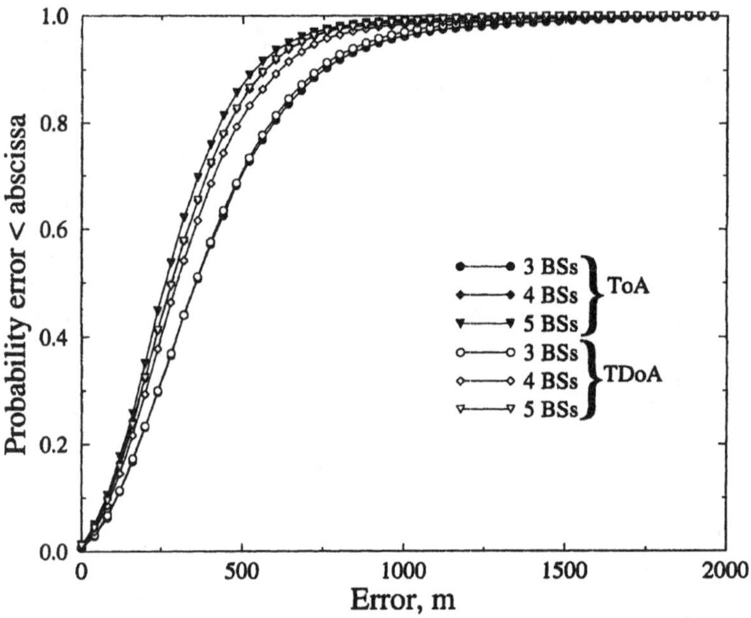

Figure 7.2. The *cdf*s of the location error resulting from the NLoS propagation for the Taylor series-based algorithm. Both ToA and TDoA results are given for various numbers of BSs used for location.

algorithm. The mean error for 4 BSs is 1200 m and for 5 BSs is 900 m. The FCC requirements are only met in 3% to 4% of the cases. Again, the larger errors result from the squaring operation that is used to form the pseudomeasurements.

Finally, we note that these results are for the geometry of the BSs and MS as specified above since the estimation bias is dependent on the geometry of the scenario according to the GDOP. When these algorithms are put in iterative form by using the KF and/or EKF, the mathematical analysis proves to be very difficult. However, we would expect those estimators to converge to near the same estimates, and thus experience approximately the same bias.

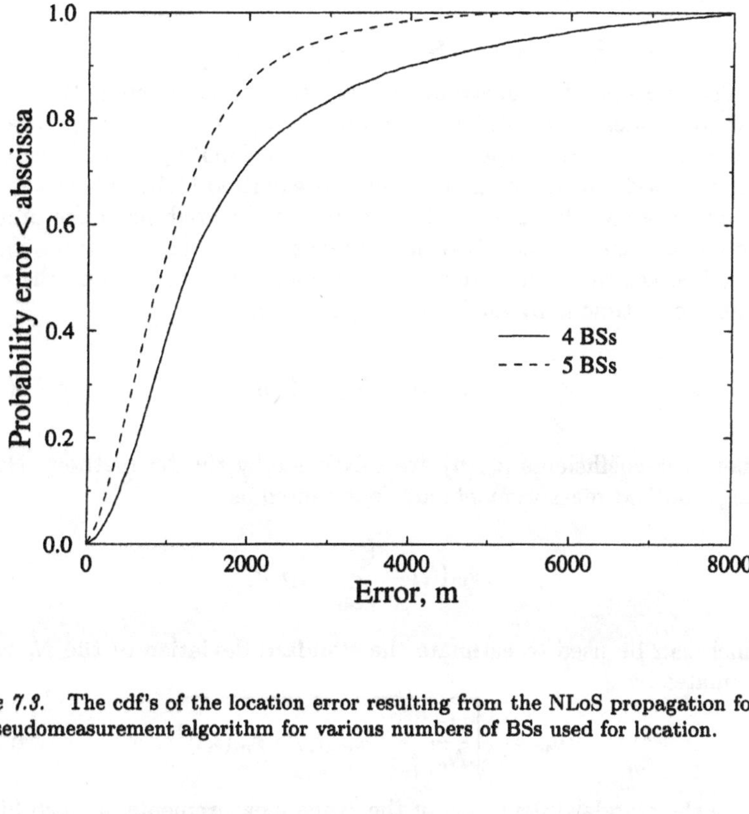

Figure 7.3. **The cdf's of the location error resulting from the NLoS propagation for the pseudomeasurement algorithm for various numbers of BSs used for location.**

2. NLOS MITIGATION TECHNIQUES

As the results of the previous section indicate, the error introduced by NLoS propagation causes a large bias in the location estimates. This impairment manifests itself in any communications system which does not maintain LoS propagation between the transmitter (MS) and receivers (BSs). In order to provide more accurate location estimates, measures must be taken to reduce or remove the effect of the bias.

Very little research can be found in the literature that addresses the problem of NLoS propagation. In this section, we discuss two methods to mitigate the error caused by NLoS propagation. The first, *LoS reconstruction* is based on the work of Wylie and Holztmann [197] which adjusts the corrupted measurements based on a time-series of range estimates. The second approach, *NLoS measurement weighting*, extends the weighting method introduced in Chapter 5 to a recursive least squares (RLS) formulation.

2.1 LOS RECONSTRUCTION

The concept of reconstructing LoS range measurements from NLoS corrupted measurements was introduced in [197]. The range measurements, obtained from the transponding of a signal by the MS back to a BS, are modeled in a similar manner to equation (7.14) where $\boldsymbol{C}(\boldsymbol{\theta}_s)$ is defined in (4.63) for ToA, and not (4.67) as the analysis of the previous section assumed. The range measurements are made at times t_i. At each BS, the range measurements are smoothed by modeling the range estimate at time t_i by an Lth-order polynomial

$$r_m(t_i) = \sum_{n=0}^{L-1} a_m(n) t_i^n \tag{7.23}$$

where the coefficients $a_m(n)$ are solved for by the LS method. Hence, the smoothed measurements are represented as

$$s_m(t_i) = \sum_{n=0}^{L-1} \hat{a}_m(n) t_i^n \tag{7.24}$$

which can be used to estimate the standard deviation of the N_r range estimates by

$$\hat{\sigma}_m = \sqrt{\frac{1}{N_r} \sum_{i=0}^{N_r-1} [s_m(t_i) - r_m(t_i)]^2} \ . \tag{7.25}$$

Since the standard deviation of the range measurements is much higher for NLoS than LoS propagation as shown by measurements in [154], $\hat{\sigma}_m$ can be used to determine if a BS is LoS or NLoS. From the data in [154], σ_m for LoS measurements is on the order of 150 m while it is approximately 409 m for NLoS-corrupted measurements. Thus, a simple hypothesis test can be used to determine the LoS and NLoS BSs based on these *a priori* statistics. To ensure that the correct decision has been made, a residual rank test was developed [197].

From knowledge of the *a priori* range error distribution at the BSs, such as that in [154], the LoS range estimates for NLoS BSs can be reconstructed. Assume that the standard measurement noise at BS i, n_i, has support on $(-\alpha_m, \alpha_m)$ and that the bias seen by BS i, b_i, has support on $(0, \beta_m)$. Location reconstruction is accomplished by first smoothing the time history of range measurements and comparing the smoothed measurements with the actual measurements. Since NLoS propagation introduces a large positive bias ($b_i > 0$) into the range estimates, the curves of the smoothed and measured ranges will be shifted up by an amount equal to the bias as compared to the true LoS ranges. The maximum deviation of the measured ranges below the smoothed ranges gives an estimate of $r_m(t_n) - \alpha_m$ at time t_n. Consequently, by shifting the smoothed range measurements down so that it passes through this point of maximum deviation and then shifting it back up by our *a priori* knowledge of α_m, the resulting smoothed curve is an estimate of the LoS ranges.

For this application to be applicable, the *a priori* statistics of range measurement errors, particularly the value of α_m, need to be known. This is dependent on the measurement hardware and methods used to estimate the ToAs. This approach depends on the prospect of having one of the raw range measurements based on a LoS signal. For the maximum deviation of the raw measurements below the smoothed measurements to be an estimate of $r_m(t_n) - \alpha_m$, the bias from NLoS propagation must be zero at time t_n. If there are no times where this occurs, the adjustment will be to a set of ranges based on the smallest bias incurred.

2.2 NLOS MEASUREMENT WEIGHTING

An alternative method to mitigate NLoS bias effects is to examine the effect that the weights of the ToA location algorithm in Chapter 5 have on performance. It was shown in that chapter that the weights can be chosen to minimize the error. The performance improvement that can be obtained by proper weighting of the NLoS corrupted measurements is illustrated in Fig. 7.4, where a three BS macrocell scenario has been assumed. The horizontal lines correspond to the case when all BSs (both LoS and NLoS) are equally weighted in the algorithm. The remaining two curves correspond to the case when the LoS BS is given unity weight, while the NLoS BS weights are varied from 0.1 to 0.9. It is seen that the performance is improved over equal weighting of the LoS and NLoS BSs. For small weights, the improvement is significant when at least one BS is LoS. If none of the BSs are LoS with the MS, then the performance is very poor regardless of the weights used. The figure indicates that improved location results are only possible when at least one of the BSs is LoS with the MS. In the following, the algorithm developed in Chapter 5 is extended to a location tracking algorithm based on the RLS method.

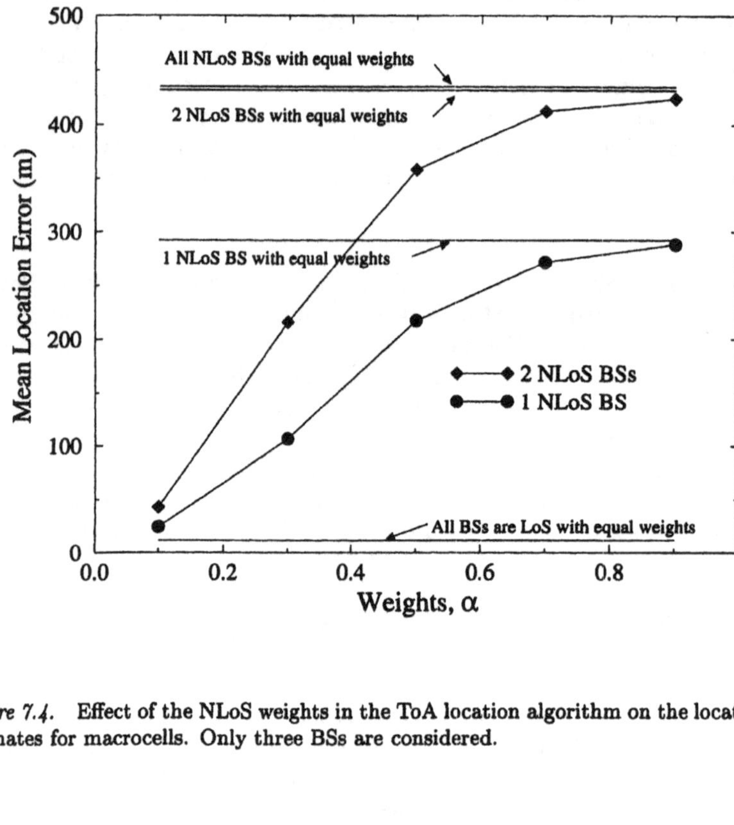

Figure 7.4. Effect of the NLoS weights in the ToA location algorithm on the location estimates for macrocells. Only three BSs are considered.

2.2.1 RECURSIVE WEIGHTING ALGORITHM

In this section, we develop the iterative extension of the NLoS weighting technique introduced in Chapter 5. Such a technique will allow the detection of NLoS BSs as the MS moves about a cell and provide a means of mitigating the effects of the NLoS biases on the location estimate. The development that follows only considers the problem of ToA location, although it can be easily generalized to the other location methods.

As in Section 1.1.2, we write the N_B noisy ToA measurements in vector form according to equation (7.3), which is repeated here for convenience,

$$r_m = C(\theta_s) + n_m \qquad (7.26)$$

where $C(\theta_s)$ is defined in equation (7.4). We define the error between the model and the measured ToAs at some measurement time i by

$$e(i;\theta_s) = r_m(i) - C(\theta_s) \ . \tag{7.27}$$

From this, we form the weighted cost function at time l with

$$\mathcal{J}(l) = \sum_{i=1}^{l}\sum_{n=1}^{N_B} \lambda^{l-i}\alpha_n e_n^2(i;\theta_s) \tag{7.28}$$

where $e_n(i;\theta_s) = [r_m]_n(i) - C_n(\theta_s)$, $[r_m]_n$ and C_n are the nth elements of r_m and $C(\theta_s)$, respectively, α_n is an element of the weighting matrix W, and λ is a "forgetting" factor which allows weighting of the time history of measurements. Throughout the rest of this chapter, the dependence of the error and measurement model on the parameter vector θ_s will be dropped for notational convenience.

The cost function in equation (7.28) can be written in two forms. The first form groups terms as follows:

$$\mathcal{J}(l) = \sum_{n=1}^{N_B} e_n^T \Lambda_n e_n \tag{7.29}$$

where we have the following definitions,

$$e_n = \begin{bmatrix} e_n(1) \\ \vdots \\ e_n(l) \end{bmatrix} \quad \text{and} \quad \Lambda_n = \begin{bmatrix} \alpha_n\lambda^{l-1} & & & 0 \\ & \ddots & & \\ & & \alpha_n\lambda & \\ 0 & & & \alpha_n \end{bmatrix} \tag{7.30}$$

whereas the second form is written as

$$\mathcal{J}(l) = \sum_{i=1}^{l} e^T(i)\Lambda(i)e(i) \tag{7.31}$$

where $e(i)$ is defined in equation (7.27) and we define

$$\Lambda(i) = \begin{bmatrix} \lambda^{l-i}\alpha_1 & & 0 \\ & \ddots & \\ 0 & & \lambda^{l-i}\alpha_{N_B} \end{bmatrix} = \lambda^{l-i}W \tag{7.32}$$

where $W = \text{diag}[\alpha_1, \ldots, \alpha_{N_B}]^T$ is a diagonal matrix of weights. The differences in the two forms depends on the grouping of the terms of the summations. In the following, the location estimator that is developed is based on the second of the forms above in equations (7.31) and (7.32).

From the these definitions and the results of the derivation in Appendix D given in equations (D.13), (D.14) and (D.15), the generalized WRLS (GWRLS) algorithm is given by the following:

$$\begin{aligned} \boldsymbol{\Gamma}_{\mathrm{RLS}}(l) &= \boldsymbol{P}(l-1)\boldsymbol{H}_C^T(l)\left[\lambda \boldsymbol{W}^{-1}\right. \\ &\qquad \left. + \boldsymbol{H}_C(l)\boldsymbol{P}(l-1)\boldsymbol{H}_C^T(l)\right]^{-1} \quad (7.33) \\ \boldsymbol{P}(l) &= \frac{1}{\lambda}\left[\boldsymbol{I} - \boldsymbol{\Gamma}_{\mathrm{RLS}}(l)\boldsymbol{H}_C(l)\right]\boldsymbol{P}(l-1) \quad (7.34) \\ \hat{\boldsymbol{\theta}}_s(l) &= \hat{\boldsymbol{\theta}}_s(l-1) + \boldsymbol{\Gamma}_{\mathrm{RLS}}(l)\left[\boldsymbol{r}_m(l) - \boldsymbol{C}(\hat{\boldsymbol{\theta}}_s(l))\right] \quad (7.35) \end{aligned}$$

where, analogous to the KF, $\boldsymbol{\Gamma}_{\mathrm{RLS}}(l)$, is the gain matrix, $\boldsymbol{P}(l)$ is the error covariance matrix and $\boldsymbol{H}_C(l)$ is the Jacobian of $\boldsymbol{C}$ defined in equation (7.9). We see that the GWRLS algorithm differs from the standard WRLS algorithm only by the presence of the $\boldsymbol{W}^{-1}$ term in equation (D.13).

2.2.2 WEIGHT ADJUSTMENT METHOD

The usefulness of the GWRLS algorithm is its ability to weight the error vector in order to minimize the impact the NLoS BSs have on the final location estimate. We consider a MS in motion such that the NLoS bias statistics over the duration of travel is accurately approximated by the measured statistics calculated by Silventoinen and Rantalainen [154]. In [154], it was determined that the standard deviation of the ToA measurements are much higher for a NLoS BS than for a LoS BS. The distribution of the bias for a NLoS BS had support on the interval [0 m, 1300 m] with a mean of approximately 500 m and a standard deviation of approximately 436 m. For a LoS BS, the mean ToA measurement error was zero with a standard deviation of 150 m. However, for these statistics to be useful, a MS must not be stationary, as they are the average statistics for the cell.

By monitoring the standard deviation of the ToA measurement vector, $\boldsymbol{r}_m$, it is possible to determine which BSs are NLoS. The weighting matrix $\boldsymbol{W}$ can then be modified to reduce the effect the NLoS BSs have on the final location estimate. A simple hypothesis test may be employed such that a BS is declared to be LoS if the standard deviation of its ToA measurements is less than 300 m, while it is declared NLoS if the standard deviation of its ToA measurements is greater than 300 m. Monitoring the standard deviation of the ToA measurements also allows for adaptation of the weights to account for BSs changing between NLoS and LoS. A windowing method can be used for monitoring purposes.

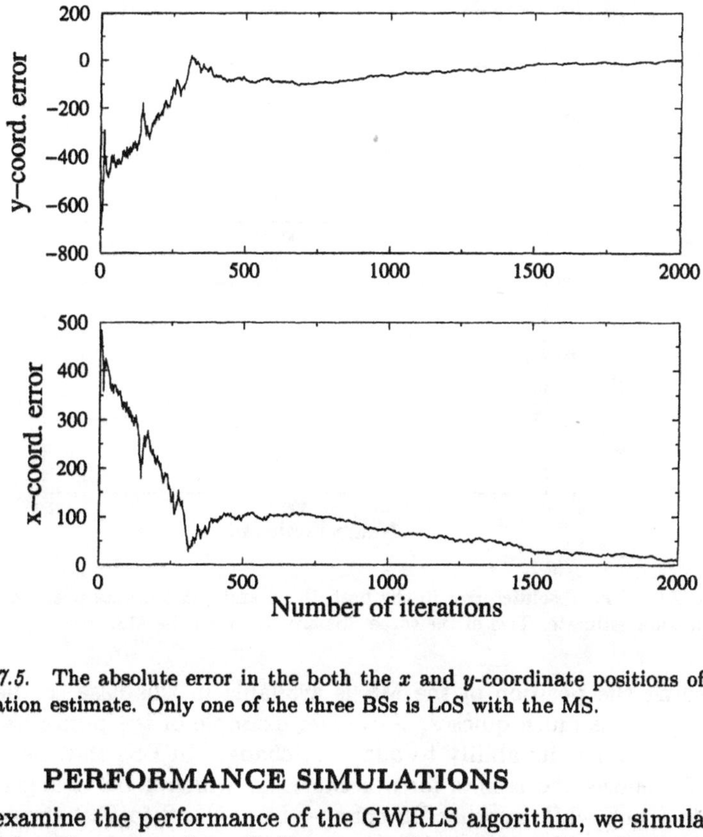

Figure 7.5. The absolute error in the both the x and y-coordinate positions of the MS location estimate. Only one of the three BSs is LoS with the MS.

2.2.3 PERFORMANCE SIMULATIONS

To examine the performance of the GWRLS algorithm, we simulated several location scenarios. Only macrocells were considered using the deployment of Fig. 7.1. The weights of the NLoS BSs, once determined, were chosen to be 0.001, while keeping the weights for the LoS BSs at 1.0. The distribution of the biases for each BS were chosen according to the distributions discussed in the previous section. The standard deviation of the ToA measurements were monitored by calculating the standard deviation over a the previous 100 measurements. Hence, a window of length 100 was used.

The ToA errors caused by the NLoS bias at each BS was modeled according to the distribution discussed in the previous section. For a three BS scenario having only a single BS with a LoS path to the MS, the results are plotted in Fig. 7.5. The figure shows that the algorithm is capable of correctly estimating the position of the MS, even if the convergence is a little slow. Fig. 7.6 shows the results for a scenario where two BSs are LoS with the MS. Obviously, more accurate information

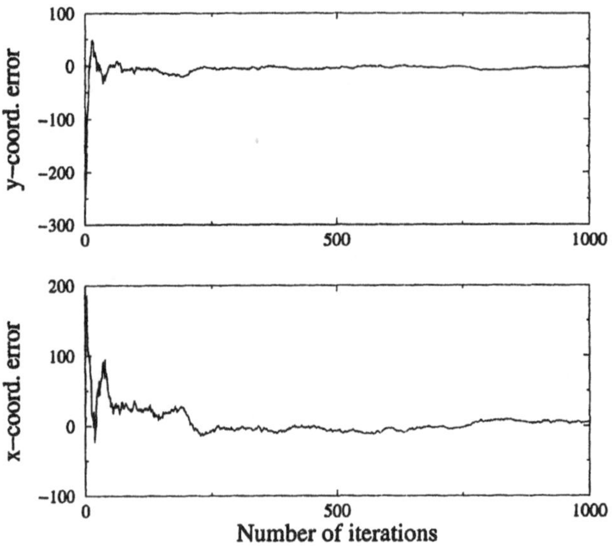

Figure 7.6. The absolute error in the both the x and y-coordinate positions of the MS location estimate. Two of the three BSs are LoS with the MS.

regarding the position of the MS is available in this case so that the convergence is much quicker. A further example of the performance of the algorithm is its ability to adapt to changes in BSs that have LoS. Fig. 7.7 shows the results for a scenario in which a BS changes from LoS to NLoS while another BS changes from NLoS to LoS abruptly at the 2000th iteration. While it takes a small amount of time in order for the filter to determine the change from the standard deviation of the ToA measurements, the algorithm quickly converges back near the true position of the MS.

3. SUMMARY

In this chapter, we have examined the effect that NLoS propagation has on two popular location algorithms: LS using a Taylor series expansion (for ToA and TDoA) and the pseudomeasurements algorithm (TDoA only). Expressions for the location estimation bias that results were derived and simulations showed that the results are quite drastic. Two methods to mitigate the effect of NLoS propagation were presented. One attempts to reconstruct the LoS measurements from the NLoS corrupted measurements while the other uses algorithm weights to reduce the contribution of the NLoS measurements to the final location solu-

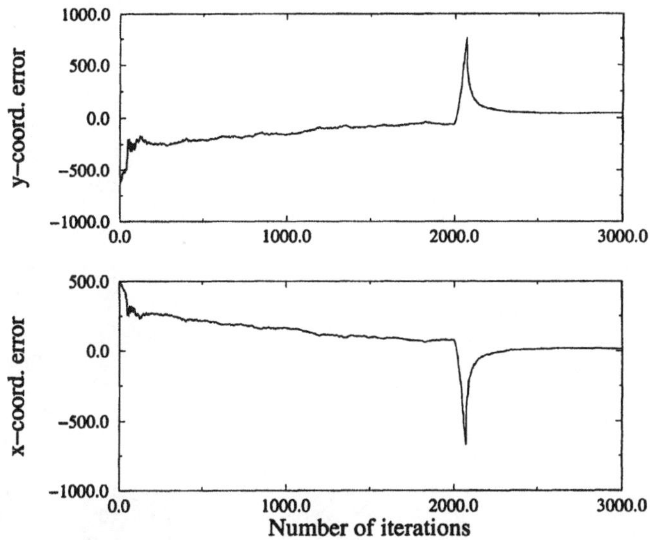

Figure 7.7. The absolute error in the both the x and y-coordinate positions of the MS location estimate. The BS that is LoS with the MS changes after 2000 iterations.

tion. Both approaches have their advantages and disadvantages. It is clear the further work needs to be done in this area to produce accurate location estimators for cellular radio systems.

Chapter 8

PROVISIONS FOR LOCATION IN WIRELESS STANDARDS

In this chapter, we will briefly discuss some of the capabilities that are available for providing location in present cellular networks.

The location requirements set forth by the FCC must be met not only by the new digital cellular systems, but the older analog system as well. With several different wireless systems on the market (AMPS, IS-54/136 TDMA, GSM, IS-95 CDMA), different methods may be necessary to implement location services in each of those systems. The signal strength method is often not implemented for cellular systems because of the large variability of received signal strength resulting from shadowing and multipath fading (see Chapter 3). AoA requires the placement of antenna arrays at the BSs which may be extremely costly. The AoA measurements can be obtained from array signal processing and is not dependent on the type of cellular system deployed. Unlike AoA, the ToA and TDoA methods require that timing information be obtained from the signals transmitted by a MS which may be implemented in different ways for each cellular system. These two methods may also require strict synchronization of the BSs. The remainder of this chapter discusses implementation strategies for the ToA and TDoA location methods in current and future generation cellular systems.

The most straightforward approach for obtaining timing information for ToA or TDoA location is the use of signal correlation methods. Specifically, maximizing cross-correlations between the signals received at pairs of BSs will provide an estimate of the TDoAs for each pair of BSs. In the following sections, we will provide a general overview of the methods by which location systems can be implemented in current cellular and PCS systems.

1. AMPS

The MS handsets in the North American analog communication systems, AMPS, have no capability to provide estimates of location parameters for positioning. Consequently, the system must rely on location using reverse link signaling. The methods that are applicable in this case are exactly the methods that have been discussed in previous chapters. The two reverse channels used by AMPS are very different in structure and in the advantages they offer for location. One method by Trueposition^{TM} of the Associated Group, Inc., uses the reverse control channel for a TDoA location system [170]. This approach is useful since the analog cellular telephones transmit on the reverse control channel even when not engaged in a voice conversation. Unfortunately,the reverse control channel cannot be used to locate a cellular phone that is engaged in a conversation. The call setup messages are sent on the control channel, but handoff messages are sent on the voice channel to conserve radio resources. Consequently, the TDoA system must include the use of the reverse voice channel as a secondary means of location.

2. CDMA

For CDMA, different methods can be used for the reverse link and forward link. On the reverse link, the timing information for ToA or TDoA can be obtained using the techniques discussed in Chapter 3. Since the BSs in IS-95 are synchronized to a GPS time reference, the time of detection of the signal from the MS can serve as a time-stamp for the time of arrival. Similarly, segments of the detected signal can be sent to a central processing office for cross-correlation in order to determine the set of TDoAs for the BSs, or they can be formed by differencing the ToAs. The signals for the ToA/TDoA measurements can come from the reverse traffic channel or the access channel. The reverse traffic channel could be used for E-911 calls, for example, since a voice call must be initially made. For other location applications, the location may be desired when the MS is not actively transmitting. In these cases, the MS could be prompted to transmit messages on the access channel in response to commands from its serving BS on the paging channels. Unfortunately, it may be not impossible to detect the MS transmissions at other BSs due to the near-far effect, or simply because the other BSs cannot "hear" them. In this case, it may be possible to have the MS power up to its maximum power for a short time. However, the use of the power up function must be limited to emergencies (such as E-911) in order to avoid excessive interference to other users. Studies have indicated that if the number of power up events are managed properly, the impact on the CDMA system is not catastrophic [53]. Also, if the number of power up events within the same cell occurs frequently, then a simple interference canceler at the BS can mitigate the interference effect.

An alternative for location in CDMA is to utilize pilot monitoring in the MS on the forward link. To assist in the handoff process, the MS monitors the strongest pilots from the surrounding BSs. The serving BS can send a pilot measurement request order (PMRO) causing the BS to respond with a message which includes the magnitudes of the pilots in the candidate set as well as the code phase of each pilot relative to its serving BS [179]. Hence, it is possible to construct TDoA estimates from these system messages. The accuracy of the TDoA estimates is dependent on the resolution of the code phase and the synchronization of the BSs. Fortunately, for IS-95, the BSs are synchronized to a GPS time reference. However, the code phase resolution is limited to a chip time, T_c, which implies a TDoAresolution (in distance) of approximately 244 m.

Finally, the soft handoff, during which the MS communicates with nearby BSs during a handoff, can be used for location in CDMA systems as long as at least three BSs are in a soft handoff with the MS. However, this approach does not provide the coverage necessary to comply with the FCC requirements since it would only be available when a MS is near a cell boundary.

3. TDMA

The TDMA based systems also provide timing information in their system messages that can be used for ToA or TDoA location. The time alignment parameter in IS-54/136 and timing advance in GSM (both abbreviated TA) are used by each of those networks to ensure that the transmissions of MSs arrive at their serving BSs in the appropriate time slots. Each BS sends the MSs a TA value which is the amount the MS must advance or retard the timing of its transmissions. Additionally, the TA serves as a measure of the propagation time between the MS and BS. The use of TAs can be applied by using forced handovers. When the location of the MS is needed, the network will force the MS to make a handover attempt from its serving BS to one of its neighboring BSs. Once the handoff request is rejected and the MS informs its serving BS of the rejection, a handoff to another BS can be attempted. This process can be continued until a sufficient number of TA measurements have been collected for location.

A primary consideration, then, is the accuracy of the TA. For IS-54, the timing of MS transmissions are advanced or retarded in units of $T_b/2$, where $T_b = 20.6\mu s$ is the bit duration. Hence, the TAs are accurate to $T_b/4$, or 1543 m. For GSM, the TA measurements are reported in units of bits, with $T_b = 3.7\mu s$. This gives a TA resolution of $T_b/2$, or 554 m, in GSM.

An alternative for GSM is to use the observed time difference (OTD) measurements which are made at the MS without forcing additional handoffs. The OTDs are used to facilitate handoffs by estimating the amount the timing of the MS would have to be advanced or retarded if it were to be handed over to another BS. With a synchronized network, the OTDs could be used to implement a TDoA location system. Unfortunately, the GSM standard does not require that the network be synchronized. Thus, some method of BS synchronism or knowledge of the timing offsets between the BS is necessary. Additionally, the OTD measurements are made to the same accuracy as the TA measurements, 554 m.

4. SUMMARY

Both CDMA and TDMA-based cellular radio systems offer the possibility of MS location built into the system. However, the resolution that can be achieved based on system messages such as the PMRO, TA or OTD are not adequate for the FCC requirements. Because of the high chip rate and the nice correlation properties of the spreading code sequences used in CDMA systems, they currently provide greater potential than the other systems for accurate location estimates. It is apparent that the resolution of the timing parameters in the system messages needs to be improved in order to provide more accurate estimates of location. Improving the accuracy of the PMRO and OTD measurements requires modification of the MS handset since those parameters are measured in the MS. When using TAs for ToA location, improvements only need to be made at the BSs. For more information, interested readers are referred to [154] and several articles in [74].

Chapter 9

CONCLUSIONS & FUTURE DIRECTIONS

The purpose of this book was to provide an overview of applications and technologies for wireless location in cellular networks, particularly CDMA systems. Using the method of radiolocation, the performance of wireless location in CDMA cellular radio systems was evaluated and possible solutions to the major impairments of those systems were presented. Performance simulations were completed which showed the effect of the number of BSs used for location and the propagation conditions on the location system. Multipath propagation, MAI and NLoS propagation were identified as the major sources of error. The effect of MAI on location systems using conventional methods for ToA estimation was analyzed and multiuser parameter estimator/trackers were discussed as possible solutions. The trackers are designed to perform well in multipath fading and near-far environments. Also, the effect of the measurement bias introduced by NLoS propagation was analyzed for two well-known location algorithms. To mitigate the effect of the bias, two approaches were discussed: one based on LoS reconstruction and the other based on reducing the influence of NLoS corrupted measurements to produce more accurate location estimates.

1. SUMMARY OF BOOK

In the following, detailed summaries are provided for the material presented in this book followed by a discussion of future work possibilities.

1.1 PERFORMANCE OF LOCATION IN CDMA SYSTEMS

Subscriber location in CDMA cellular networks was investigated for both macrocellular and microcellular deployments. For range estimation, it was seen that location error and standard deviation increases with increasing shadow standard deviation and decreasing E_c/N_o. Under conditions of moderate delay spread, the tracking performance of the DLL can be improved by using a smaller early-late discriminator offset, Δ. For 2-D location, the location error is reduced by increasing the number of BSs used in the location process. For macrocells, the unconstrained NL-LS ToA method outperforms the AoA method for a small scattering radius, while the the AoA method performs slightly better for a large scattering radius. In all cases, the constrained NL-LS ToA method performed best.

It should be noted that the AoA-based location results depend on the method chosen for generating the AoAs, namely, reflection from a ring of scatterers about the MS. A different method for generating the AoA distribution, $p(\theta_{\mathrm{AoA}})$, may produce different results. For microcells, the AoA method may be inappropriate so only the ToA method was used. Again, using more BSs decreases the error and using small weights for the NLoS BSs provides good error reduction.

The ToA method relies on a NL-LS solution. This was obtained here by using the steepest descent method, chosen for its simplicity. Better methods for locating the minimum of a least squares surface could be employed, such as the Levenberg-Marquardt method, to improve the convergence time and possibly the accuracy of the location estimates.

1.2 EFFECTS OF MULTIPLE-ACCESS INTERFERENCE ON WIRELESS LOCATION

The effects of multiple-access interference on the performance of the NC-DLL were examined. The criteria evaluated were the MTLL and the *rms* tracking error through the use of both linear and nonlinear (renewal theory) approaches. As expected, increasing the number of interferers drastically reduces the MTLL and increases the *rms* tracking error. When a single user is received with significantly more power than the desired user (near-far effect), even in a two-user scenario, the effects are worse than when there are many users with equal received strength, showing the importance of good power control algorithms. The results also show that the use of long PN code sequence lengths provides the same performance as when there are no interferers present due to reduction in the magnitude of the normalized cross-correlation between two users' sequences. Consequently, for a fixed number of users, longer codes may be utilized to improve the tracking performance of the NC-DLL. The performance gain obtained from longer spreading codes is also apparent when considering the load of the system.

The results presented in this book are based on the accurate approximation of the ACF. We have assumed that this is accurate for the family of PN sequences that are of interest (i.e., *m*-sequences, Gold sequences, etc.). It should be noted that more accurate results may be obtained by use of known cross-correlation properties of families of PN sequences. For instance, the periodic cross-correlation functions for a family of Gold codes takes on only three values with known probability [164]. This information could be used to tailor the analysis for Gold codes, but would not be applicable for other families of codes. The analysis in this book is intended to be generally applicable to all families of PN codes.

The results from the ACF approximation were compared to those obtained from the Gaussian approximation for the MAI. It was shown that the Gaussian approximation was not accurate for high SNRs, short spreading codes, when there are a few interferers, or when a single user dominates the interference. These results also confirm other studies in the literature.

To mitigate the problem of MAI, solutions based on multiple-user parameter estimators and trackers were discussed. These estimators provide the ability to accurately estimate and/or track parameters, such as time delays, accurately in MAI. We presented a parameter estimator based on the UF that is capable of estimating and tracking time-varying channel amplitudes and time delays in MAI. From former performance evaluations [82, 83], the UF was expected to have superior performance to the standard EKF. The nonlinear UF avoids the pitfalls of using the EKF since it does not approximate the nonlinear problem with a linear one. It also avoids the calculation of Jacobians, which are required for the EKF. The UF-based estimator was shown to have the ability to converge to the users' true amplitudes and time delays for equal powered users and a near-far ratio of 20 dB. It also was capable of tracking time varying delays in a near-far environment. For multipath fading channels, the estimator was able to accurately track the time-varying channel magnitudes which were Rayleigh faded with different fade rates. Although the users in the fading channel simulations experienced extremes in the instantaneous near-far ratio, accurate tracking was still maintained.

Finally, the MSE of the amplitude and delay estimates were compared to the CRLB. The lack of dependence of the CRLB for each user on parameters of the other users indicates that may be possible to find an estimator that is near-far resistant. The MSE from the simulations is much higher than the CRLB, with the near-far situations having slightly higher MSE then when users are perfectly power-controlled. The MSE is seen to decrease as the SNR is increased and as the number of samples that are used for convergence is increased. The results also indicate that as the number of samples is increased, the MSE for near-far situations begins to approach that of the power-controlled situation.

1.3 EFFECTS OF NON-LINE-OF-SIGHT PROPAGATION ON WIRELESS LOCATION

The effect of the error introduced by NLoS propagation was investigated for different location algorithms. The location estimate bias was derived for ToA and TDoA location estimators based on a Taylor series approximation to the measurement model. The bias was also derived for the pseudomeasurement algorithm. Location tracking algorithms, such as the EKF and RLS are not analyzed due to the difficult nature of such a task. However, it is noted that these algorithms should converge to near the same result as the estimation algorithms. The NLoS error is shown to introduce a bias that leads to large errors in the location estimates.

Two approaches for mitigating the NLoS problem were discussed. The first was based on reconstructing the LoS range measurements in a ToA location system by using *a priori* knowledge of measurement error statistics. The second approach evaluated mitigating the NLoS bias that corrupts the ToA measurements by using algorithmic weights. As the results of Chapter 5 indicated, it was possible to reduce the effect of the NLoS bias by weighting the contribution of the NLoS BSs to the overall location estimate. The technique was based on generalized WRLS filtering which was able to locate the position of the MS within the requirements set forth by the FCC. The method was also capable of accounting for changing NLoS BSs. Accurate location estimates could only be obtained if at least one BS (most likely the serving BS) was LoS with the MS. Also, the technique is only applicable to a moving MS in order to make use of the statistics of range errors that have been measured for a cell.

2. SUGGESTIONS FOR FUTURE RESEARCH

Several areas of further research are possible. These mainly address methods to mitigate the major impairments to accurate location. The analyses and solutions to the major impairments presented in this book are only a starting point.

2.1 MULTIPATH MITIGATION TECHNIQUES

Methods for improving the accuracy of the ToA measurements in a multipath fading environment have not been addressed in this book in order to give more attention to MAI and NLoS propagation impairments. Multipath propagation introduces errors in the ToA measurements. Several approaches can be investigated including frequency domain, super-resolution, and other techniques that can identify closely spaced multipaths.

2.2 MAI MITIGATION TECHNIQUES

Although MAI affects all forms of wireless communications systems, it is particularly a problem for CDMA systems since all users share the same bandwidth and are only separated by the use of PN spreading codes. Multiuser detectors and parameter estimators have been studied by many investigators to allow acceptable performance in multiple user systems. Most methods are too complex for practical implementation so alternative methods are necessary to produce near-optimal performance with a minimum amount of complexity. For the radiolocation problem, parameter estimators, as opposed to detectors only, that are near-far resistant are the main concern.

2.3 NLOS MITIGATION TECHNIQUES

The area that requires the most amount of attention is the problem of NLoS propagation. The biased measurements that are produced in such an environment are not unique to CDMA systems. All wireless communications systems face the NLoS problem, unless they are specifically designed to maintain LoS between the transmitter and receiver(s) such as GPS. Unless the NLoS bias is accounted for in some way, the location estimates will be far from the true location, regardless of the accuracy to which ToA estimates can be made in multipath fading channels and MAI.

A possible approach is to estimate the bias along with the location of the MS in order to remove its effect on the measurements. With this method, the number of unknowns is increased by the length of the bias vector, which is equal to the number of BSs used for location. Unfortunately, both the state augmentation approach, where the bias vector is included in the parameter vector to be estimated, and the adaptive Kalman filtering[1] approach, which uses multiple KFs each designed around a particular value of the bias vector, suffer from a lack of observability of the unknown parameters. Consequently, other approaches must be developed, perhaps improving the observability of the unknowns by adding additional measurements or information to the system.

2.4 FORWARD LINK LOCATION

The investigations in this book have focused on reverse link radiolocation where the radio signals are received and processed at the BSs. An alternative is to apply the same location approaches on the forward link such that the ToAs (or TDoAs) are measured at the MS. In IS-95 CDMA, it is possible to use the MS's measurements of nearby BSs' pilot signals to form a location estimate [137]. The pilot measurements are made by the MS to assist in the handoff process. The serving BS can send a pilot measurement request order (PMRO) causing the BS to respond with a message which includes the magnitudes of the pilots in the candidate set of BSs as well as the code phase of each BS pilot relative to its serving BS [179]. Hence, it is possible to construct TDoA estimates from these system messages. The accuracy of the TDoA estimates is dependent on the resolution of the code phase and the synchronization of the BSs. Fortunately, for IS-95, the BSs are synchronized to a GPS time reference. However, the code phase resolution is limited to a chip time, T_c, which implies a TDoA resolution of approximately 244 m.

Unfortunately, the location estimates obtained with this approach are not very accurate since the resolution of the pilot measurements is limited to one chip duration. As the CDMA standard evolves, requirements for improved resolution of the pilot measurements could be instituted, thus improving the accuracy of the location estimates. For future generation wireless systems, such as International Mobile Telecommunications 2000 (IMT-2000) and Universal Mobile Telecommunications System (UMTS), the opportunity presently exists to have the necessary location functionality specified in the emerging standards. For instance, accurate pilot signal measurements can be required in the standard making forward link location based in the MS possible, without the need of an overlay system.

[1] Adaptive Kalman filtering is also referred to as the *multiple-model* approach in the literature [2, 5, 109].

Appendix A
Derivation of $R_{n_T}(\xi, \varepsilon)$

In this appendix, the autocorrelation function, $R_{n_T}(\xi, \varepsilon)$, of the total noise, $n_T(t, \varepsilon)$, in equation (6.8) is derived. The total noise process consists of five components

$$n_T(t, \varepsilon) = n_1(t) + n_2(t, \varepsilon) + n_3(t, \varepsilon) + n_4(t, \varepsilon) + n_5(t, \varepsilon) \qquad \text{(A.1)}$$

where the individual components are given in equations (6.9)-(6.13). It is easy to show that the noise processes $n_i(t, \varepsilon)$ are pairwise uncorrelated and, therefore,

$$R_{n_T}(\xi, \varepsilon) = R_{n_1}(\xi) + R_{n_2}(\xi, \varepsilon) + R_{n_3}(\xi, \varepsilon) + R_{n_4}(\xi, \varepsilon) + R_{n_5}(\xi, \varepsilon) \quad \text{(A.2)}$$

where $R_{n_i}(\xi, \varepsilon) = \mathrm{E}[n_i(t, \varepsilon) n_i(t + \xi, \varepsilon)]$, $i = 1, \ldots, 5$. Proceeding with each term of (A.2) individually, the first and second terms can be expressed as [130]

$$\begin{aligned} R_{n_1}(\xi) &= 2N_0^2 \left[1 - R_a^2(2\Delta)\right] \left[\int_{-\infty}^{\infty} |H_l(f)|^2 e^{j2\pi f \xi} \, df\right]^2 && \text{(A.3)} \\ R_{n_2}(\xi, \varepsilon) &= 2P_1 N_0 R_{\bar{d}}(\xi) \left[R_a^2(\varepsilon - \Delta) + R_a^2(\varepsilon + \Delta) \right. \\ &\quad \left. -2R_a(2\Delta) R_a(\varepsilon - \Delta) R_a(\varepsilon + \Delta)\right] \\ &\quad \cdot \int_{-\infty}^{\infty} |H_l(f)|^2 e^{j2\pi f \xi} \, df && \text{(A.4)} \end{aligned}$$

where

$$R_{\bar{d}}(\xi) = \mathrm{E}\left[\bar{d}_1(t + \xi) \bar{d}_1(t)\right] . \qquad \text{(A.5)}$$

Evaluating $R_{\bar{d}}(\xi)$ at $\xi = 0$ gives D_2 of equation (6.14).

The derivations of the autocorrelation functions $R_{n_i}(\xi, \varepsilon), i = 3, \ldots, 5$, are more complicated due to the presence of the cross-correlation terms

$R_{k,1}(\eta)$ in the noise expressions, which we treat as random variables as discussed in Chapter 6. In forming the autocorrelation functions of these noise processes, terms of the form $\mathrm{E}[R_{k,1}(\eta)R_{j,1}(\nu)]$ need to be evaluated. Taking $l_\eta = \lfloor \eta \rfloor$ and $l_\nu = \lfloor \nu \rfloor$ with $s_\eta = \eta - l_\eta$ and $s_\nu = \nu - l_\nu$, $R_{k,1}(\eta)$ and $R_{j,1}(\nu)$ can be expressed as a random variable in terms of the ACFs $C_{k,1}(l_\eta)$ and $C_{j,1}(l_\nu)$, respectively. For this analysis, we need to examine two cases: $k \neq j$ and $k = j$.

<u>Case 1:</u> $k \neq j$
Assuming that the cross-correlation between two users' PN sequences is independent of the cross-correlation of another two users' PN sequences, we have $\mathrm{E}[R_{k,1}(\eta)R_{j,1}(\nu)] = 0$.

<u>Case 2:</u> $k = j$
From the noise process definitions $n_3(t, \varepsilon)$ through $n_5(t, \varepsilon)$, we see that η and ν will be of the form $\tau_d^{(k)} + \varepsilon \pm \Delta$. First, we will consider the case where $\eta = \nu = \tau_d^{(k)} + \varepsilon \pm \Delta$, which implies that $l_\eta = l_\nu$ and $s_\eta = s_\nu$. Using the definitions above, the expected value of the cross-correlation product can be shown to be, for $\eta = \nu$:

$$\mathrm{E}[R_{k,1}(\eta)R_{k,1}(\nu)] = \frac{2}{9N^2}\Lambda(\Lambda + 2) \tag{A.6}$$

where $\Lambda = \lfloor \sqrt{3N} - 1 \rfloor / N$. Next, we consider the case where $\eta \neq \nu$, namely $\eta = \tau_d^{(k)} + \varepsilon \pm \Delta$ and $\nu = \tau_d^{(k)} + \varepsilon \mp \Delta$, such that $\nu = \eta \pm 2\Delta$. Without loss of generality, assume that $\eta = \tau_d^{(k)} + \varepsilon + \Delta$ and $\nu = \tau_d^{(k)} + \varepsilon - \Delta$. Using l_η, l_ν, s_η, and s_ν as defined above, only three cases are possible for $0 < \Delta < 1$: $l_\eta = l_\nu$, $l_\eta = l_\nu + 1$, and $l_\eta = l_\nu + 2$. For the first case, $l_\eta = l_\nu$ implies that $s_\eta = s_\nu + 2\Delta$. In the second case, $l_\eta = l_\nu + 1$ implies that $s_\eta = s_\nu + 2\Delta - 1$. For the final case, $l_\eta = l_\nu + 2$ implies that $s_\eta = s_\nu + 2\Delta - 2$. For these cases and using the assumption that $\mathrm{E}[C_{k,1}(l)C_{k,1}(l + \alpha)] = 0$ for $\alpha \neq 2mN$, m an integer, we have for $\eta \neq \nu$:

$$\mathrm{E}[R_{k,1}(\eta)R_{k,1}(\nu)] = \begin{cases} \frac{2}{9N^2}\Lambda(\Lambda+2), & l_\eta = l_\nu \\ \frac{1}{3N^2}(\frac{2}{3} - \Delta)\Lambda(\Lambda+2), & l_\eta = l_\nu + 1 \\ 0, & l_\eta = l_\nu + 2 \end{cases} . \tag{A.7}$$

The assumption that the random variable $C_{k,i}(l)$ is uncorrelated from non-zero shifts of itself can be validated by examining its autocorrelation function. Figs. A.1 and A.2 show a typical autocorrelation function and an average autocorrelation function of the ACF, respectively, for the family of Gold codes formed from the preferred pair of m-sequences of

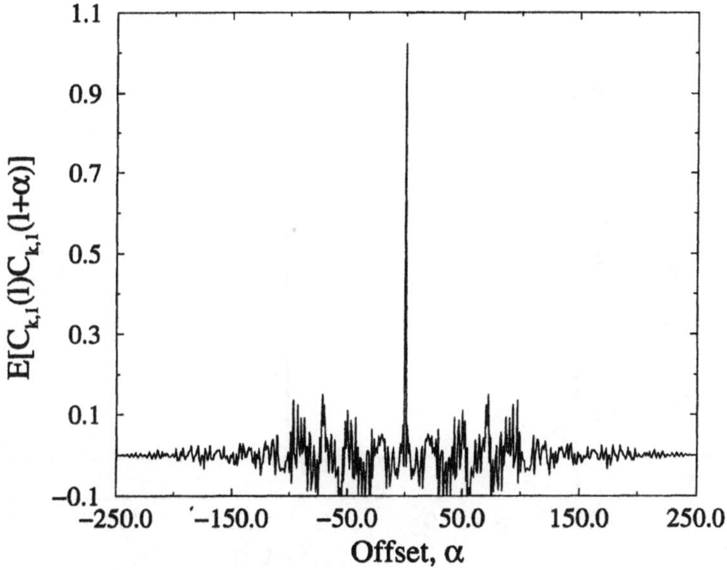

Figure A.1. A typical autocorrelation function of the ACF for a pair of codes in a family of Gold codes of length 27.

length 127 generated by the polynomials x^7+x^3+1 and $x^7+x^3+x^2+x+1$.

The average autocorrelation function is formed by averaging over all correlation functions in the family of codes formed by the preferred pair above. We see the autocorrelation is small for offsets other than zero; hence, the assumption is valid. We note that $\mathrm{E}[R_{k,1}(\eta)R_{j,1}(\nu)]$ depends on the values of the random variables l_η and l_ν. To find their effect on this quantity, we need to average over their discrete distributions. Hence, for $j = k$, we need to determine the probabilities for $l_\eta = l_\nu$, $l_\eta = l_\nu + 1$, and $l_\eta = l_\nu + 2$. These quantities are expressed as:

$$\begin{aligned} P(l_\eta = l_\nu) &= \mathrm{Prob}\Big[l_\nu \leq (\tau_d^{(k)} + \varepsilon - \Delta) \leq l_\nu + 1, \\ &\qquad\qquad l_\nu \leq (\tau_d^{(k)} + \varepsilon + \Delta) \leq l_\nu + 1\Big] \\ &= \mathrm{Prob}\Big[l_\nu - \varepsilon + \Delta \leq \tau_d^{(k)} \leq l_\nu + 1 - \varepsilon - \Delta\Big] \\ &= \begin{cases} 1 - 2\Delta & 0 \leq \Delta < \frac{1}{2} \\ 0 & \frac{1}{2} \leq \Delta \leq 1 \end{cases} \end{aligned} \tag{A.8}$$

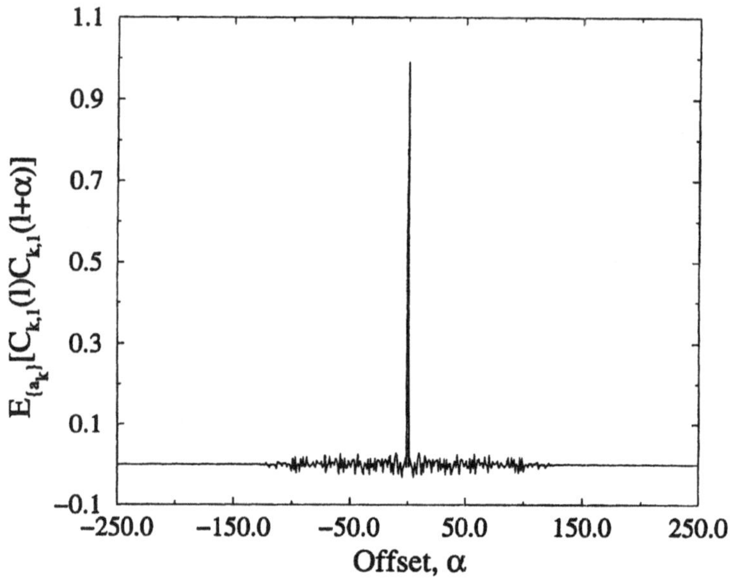

Figure A.2. The average autocorrelation function of the ACF for all codes in a family of Gold codes of length 127.

where we have used the fact that $\tau_d^{(k)}$ is uniformly distributed on $[0, N)$. Following a similar procedure, we have

$$\begin{aligned} P(l_\eta = l_\nu + 1) &= \text{Prob}\Big[l_\nu \le (\tau_d^{(k)} + \varepsilon - \Delta) \le l_\nu + 1, \\ &\qquad l_\nu + 1 \le (\tau_d^{(d)} + \varepsilon + \Delta) \le l_\nu + 2\Big] \\ &= \begin{cases} 2\Delta & 0 \le \Delta < \frac{1}{2} \\ 2(1-\Delta) & \frac{1}{2} \le \Delta \le 1 \end{cases} \end{aligned} \tag{A.9}$$

and

$$\begin{aligned} P(l_\eta = l_\nu + 2) &= 1 - P(l_\eta = l_\nu) - P(l_\eta = l_\nu + 1) \\ &= \begin{cases} 0 & 0 \le \Delta < \frac{1}{2} \\ 2\Delta - 1 & \frac{1}{2} \le \Delta \le 1 \end{cases} \end{aligned} \tag{A.10}$$

Since l_η and l_ν are discrete random variables, then the autocorrelation functions $R_{n_i}(\xi)$ $(i = 3, 4, 5)$, are given by

$$R_{n_i}(\xi) = R_{n_i}(\xi | l_\eta = l_\nu)P(l_\eta = l_\nu) + R_{n_i}(\xi | l_\eta = l_\nu + 1)P(l_\eta = l_\nu + 1)$$

$$+R_{n_i}(\xi | l_\eta = l_\nu + 2) P(l_\eta = l_\nu + 2) \qquad \text{(A.11)}$$

Using the above analysis, the autocorrelation functions for $n_3(t,\varepsilon)$, $n_4(t,\varepsilon)$, and $n_5(t,\varepsilon)$ can be shown to be:

$$R_{n_3}(\xi,\varepsilon) = \frac{4}{9N^2} N_0 R_{\tilde{d}}(\xi) \Lambda(\Lambda+2) v_1(\Delta) \left[\sum_{k=2}^{K} P_k\right] \cdot \int_{-\infty}^{\infty} |H_l(f)|^2 e^{j2\pi f\xi}\, df \qquad \text{(A.12)}$$

$$R_{n_4}(\xi,\varepsilon) = \frac{4}{9N^2} \Lambda(\Lambda+2) P_1 R_{\tilde{d}}^2(\xi) \left[\sum_{k=2}^{K} P_k\right] \left[R_a^2(\varepsilon-\Delta) + R_a^2(\varepsilon+\Delta) - 2v_2(\Delta) R_a(\varepsilon-\Delta) R_a(\varepsilon+\Delta)\right] \qquad \text{(A.13)}$$

$$\begin{aligned} R_{n_5}(\xi,\varepsilon) = {} & \frac{2}{225N^4} \Lambda(\Lambda+2) R_{\tilde{d}^2}(\xi) \left[\sum_{k=2}^{K} P_k^2\right] \cdot \left[32\Lambda^2 + 64\Lambda - 6 + w_1(\Delta)(9\Lambda^2 + 18\Lambda - 2)\right] \\ & + \frac{4}{81N^4} \Lambda^2(\Lambda+2)^2 R_{\tilde{d}}^2(\xi) w_2(\Delta) \left[\sum_{k=2}^{K} \sum_{\substack{j=2 \\ j\neq k}}^{K} P_k P_j\right] \end{aligned} \qquad \text{(A.14)}$$

where

$$R_{\tilde{d}^2}(\xi) = \mathrm{E}\left[\tilde{d}^2(t+\xi)\tilde{d}^2(t)\right] . \qquad \text{(A.15)}$$

The functions $v_1(\Delta)$, $v_2(\Delta)$, $w_1(\Delta)$, and $w_2(\Delta)$ are defined as follows:

$$\begin{aligned} v_1(\Delta) &= \begin{cases} 1 - (1-3\Delta^2) R_a(2\Delta) & 0 \le \Delta \le \frac{1}{2} \\ 1 - (1-\Delta)(2-3\Delta) R_a(2\Delta) & \frac{1}{2} \le \Delta \le 1 \end{cases} \\ w_1(\Delta) &= \begin{cases} -(1-\Delta)(20\Delta^2 - 25\Delta + 8) & 0 \le \Delta \le \frac{1}{2} \\ 20\Delta^3 + 5\Delta^2 - 2\Delta - 3 & \frac{1}{2} \le \Delta \le 1 \end{cases} \\ v_2(\Delta) &= \begin{cases} 1 - 3\Delta^2 & 0 \le \Delta \le \frac{1}{2} \\ 1 - 3\Delta + 3\Delta^2 & \frac{1}{2} \le \Delta \le 1 \end{cases} \\ w_2(\Delta) &= \begin{cases} -(9\Delta^4 - 30\Delta^3 + 37\Delta^2 - 20\Delta + 3) & 0 \le \Delta \le \frac{1}{2} \\ 3\Delta^2(2 - 3\Delta^2) & \frac{1}{2} \le \Delta \le 1 \end{cases} \end{aligned} \qquad \text{(A.16)}$$

Appendix B
Kalman Filter Equations

In this appendix, the equations for the EKF algorithm are given. They are useful for the development of the multiuser parameter estimator in Chapter 6 and the NLoS mitigation technique of Chapter 7.

Given the state and measurement models for $\boldsymbol{\theta}(n)$ and $z(n)$,

$$\begin{aligned} \boldsymbol{\theta}(n+1) &= \boldsymbol{\Phi}\boldsymbol{\theta}(n) + \boldsymbol{w}(n) && \text{(B.1)} \\ \boldsymbol{z}(n) &= h(\boldsymbol{\theta}(n)) + \boldsymbol{v}(n) && \text{(B.2)} \end{aligned}$$

respectively, where $\boldsymbol{\Phi}$ is the state transition matrix, $\boldsymbol{w}(n)$ is the process noise, $\boldsymbol{v}(n)$ is the measurement noise, and $h(\boldsymbol{\theta}(n))$ is the nonlinear measurement function of the unknown parameter vector $\boldsymbol{\theta}(n)$, the standard EKF provides an iterative algorithm for computing an estimate of the unknown parameter vector $\boldsymbol{\theta}(n)$. The process and measurement noise noise terms are assumed to be independent AWGN process such that

$$\begin{aligned} \mathrm{E}\left\{\boldsymbol{w}(i)\boldsymbol{w}^T(j)\right\} &= \delta_{ij}\boldsymbol{Q} && \text{(B.3)} \\ \mathrm{E}\left\{\boldsymbol{v}(i)\boldsymbol{v}^T(j)\right\} &= \delta_{ij}\boldsymbol{\Sigma} && \text{(B.4)} \\ \mathrm{E}\left\{\boldsymbol{w}(i)\boldsymbol{v}^T(j)\right\} &= \mathbf{0},\ \forall i,j && \text{(B.5)} \end{aligned}$$

where $\delta_i = 1$ for $i = j$ and $\delta_{ij} = 0$ otherwise. The linear update equations for the state and error covariance are

$$\begin{aligned} \hat{\boldsymbol{\theta}}(n+1|n+1) &= \hat{\boldsymbol{\theta}}(n+1|n) + \boldsymbol{\Gamma}_{\mathrm{KF}}(n+1)\nu(n+1) && \text{(B.6)} \\ \boldsymbol{P}(n+1|n+1) &= \boldsymbol{P}(n+1|n) \\ &\quad -\boldsymbol{\Gamma}_{\mathrm{KF}}(n+1)\boldsymbol{P}_{\nu\nu}(n+1|n)\boldsymbol{\Gamma}_{\mathrm{KF}}^T(n+1) && \text{(B.7)} \end{aligned}$$

where $\boldsymbol{\Gamma}_{\mathrm{KF}}(n+1)$ is the *Kalman gain* given by

$$\boldsymbol{\Gamma}_{\mathrm{KF}}(n+1) = \boldsymbol{P}_{\theta z}(n+1|n)\boldsymbol{P}_{\nu\nu}^{-1}(n+1|n) \qquad \text{(B.8)}$$

and $\nu(n+1) = z(n+1) - \hat{z}(n+1|n)$ is the *innovation* where $\hat{z}(n+1|n)$ is the predicted observation. The covariance of the innovation can be shown to be

$$\boldsymbol{P}_{\nu\nu}(n+1|n) = \boldsymbol{P}_{zz}(n+1|n) + \boldsymbol{\Sigma} \ . \tag{B.9}$$

The quantity $\boldsymbol{P}_{\theta z}(n+1|n)$ is the predicted cross-covariance matrix between $\hat{\boldsymbol{\theta}}(n+1|n)$ and $\hat{z}(n)$.

For the EKF, a Taylor series expansion of the nonlinear function, $h(\boldsymbol{\theta}(n))$, about the predicted estimate, $\hat{\boldsymbol{\theta}}(n+1|n)$, is truncated to first order resulting in the presence of the Jacobian, $\boldsymbol{H}$, in the filter update equations. In this formulation, the predicted observation is given by

$$\hat{z}(n+1|n) = h\left(\hat{\boldsymbol{\theta}}(n+1|n)\right) \tag{B.10}$$

with covariance

$$\boldsymbol{P}_{\nu\nu}(n+1|n) = \boldsymbol{H}\boldsymbol{P}(n+1|n)\boldsymbol{H}^T + \boldsymbol{\Sigma} \ . \tag{B.11}$$

The cross-covariance matrix $\boldsymbol{P}_{\theta z}(n+1|n)$ is found to be

$$\boldsymbol{P}_{\theta z}(n+1|n) = \boldsymbol{P}(n+1|n)\boldsymbol{H}^T \ . \tag{B.12}$$

Finally, the error covariance in evolves in time according to

$$\boldsymbol{P}(n+1|n) = \boldsymbol{\Phi}\boldsymbol{P}(n|n)\boldsymbol{\Phi}^T + \boldsymbol{Q} \ . \tag{B.13}$$

Hence, we see that the estimates produced by the EKF are made on the assumption that the error in truncating the Taylor series expansion to first order is small.

Appendix C
Derivation of the Cramér-Rao Bound

This appendix develops the Cramér-Rao bound for the signal model used in Chapter 6. The derivation of the Cramér-Rao bound is based on computing the gradient of the log-likelihood function with respect to the unknown parameter vector, $\boldsymbol{\theta}_p$, as follows:

$$\ln \Lambda(\boldsymbol{r}) = C - \frac{1}{2\sigma_\eta^2} \sum_{l=1}^{L} \left[r(l) - \sum_{k=1}^{K} \beta_k(l) d_{k,m_l} \bar{a}_k(lT_s - m_l T_b - \tau_k(l)) \right]^2 \tag{C.1}$$

where $\boldsymbol{r}$ is a $L \times 1$ vector of received signal samples and C is a constant. The gradient is defined to be

$$\frac{\partial}{\partial \boldsymbol{\theta}_p} \ln \Lambda(\boldsymbol{r}) = \begin{bmatrix} \frac{\partial}{\partial \boldsymbol{\beta}} \ln \Lambda(\boldsymbol{r}) \\ \frac{\partial}{\partial \boldsymbol{\tau}} \ln \Lambda(\boldsymbol{r}) \end{bmatrix} . \tag{C.2}$$

For each individual term, β_i, of the vector $\boldsymbol{\beta}$, the derivative of $\Lambda(\boldsymbol{r})$ with respect to β_i is given by

$$\frac{\partial}{\partial \beta_i} \ln \Lambda(\boldsymbol{r}) = -\frac{1}{\sigma_\eta^2} \sum_{l=1}^{L} \eta(l) d_{i,m_l} \bar{a}(lT_s - m_l T_b - \tau_i(l)) \tag{C.3}$$

where $\eta(l) = r(l) - \boldsymbol{\beta}(l)\boldsymbol{\mathcal{D}}(l)\bar{\boldsymbol{a}}(l)$ from equation (2.25). Likewise, the derivative of $\Lambda(\boldsymbol{r})$ with respect to τ_i is given by

$$\frac{\partial}{\partial \tau_i} \ln \Lambda(\boldsymbol{r}) = -\frac{1}{\sigma_\eta^2} \sum_{l=1}^{L} \eta(l) \beta_i(l) d_{i,m_l} \cdot \frac{\partial}{\partial \tau_i} \bar{a}_i(lT_s - m_l T_b - \tau_i(l)) . \tag{C.4}$$

From the definition in (6.77) and using (C.2), the Fisher information matrix is found to be

$$J = \begin{bmatrix} J_{\beta\beta} & J_{\beta\tau} \\ J_{\beta\tau}^T & J_{\tau\tau} \end{bmatrix} \tag{C.5}$$

where the $K \times K$ matrices of the four blocks are given by

$$J_{\beta\beta} = \mathrm{E}\left\{\left[\frac{\partial}{\partial\beta}\ln\Lambda(r)\right]\left[\frac{\partial}{\partial\beta}\ln\Lambda(r)\right]^T\right\} \tag{C.6}$$

$$J_{\beta\tau} = \mathrm{E}\left\{\left[\frac{\partial}{\partial\beta}\ln\Lambda(r)\right]\left[\frac{\partial}{\partial\tau}\ln\Lambda(r)\right]^T\right\} \tag{C.7}$$

$$J_{\tau\tau} = \mathrm{E}\left\{\left[\frac{\partial}{\partial\tau}\ln\Lambda(r)\right]\left[\frac{\partial}{\partial\tau}\ln\Lambda(r)\right]^T\right\}. \tag{C.8}$$

Using these definitions, we can identify the ij^{th} element of each of these matrices as follows:

$$\begin{aligned}\left[J_{\beta\beta}\right]_{ij} &= \mathrm{E}\left\{\left[\frac{\partial}{\partial\beta_i}\ln\Lambda(r)\right]\left[\frac{\partial}{\partial\beta_j}\ln\Lambda(r)\right]\right\} \\ &= \frac{1}{\sigma_\eta^2}\sum_{l=1}^{L} d_{i,m_l} d_{j,m_l}\bar{a}_i(lT_s - m_lT_b - \tau_i(l)) \\ &\qquad \cdot\bar{a}_j(lT_s - m_lT_b - \tau_j(l))\end{aligned} \tag{C.9}$$

$$\begin{aligned}\left[J_{\beta\tau}\right]_{ij} &= \mathrm{E}\left\{\left[\frac{\partial}{\partial\beta_i}\ln\Lambda(r)\right]\left[\frac{\partial}{\partial\tau_j}\ln\Lambda(r)\right]\right\} \\ &= \frac{1}{\sigma_\eta^2}\sum_{l=1}^{L} d_{i,m_l} d_{j,m_l}\beta_j(l)\bar{a}_i(lT_s - m_lT_b - \tau_i(l)) \\ &\qquad \cdot\bar{a}_j^d(lT_s - m_lT_b - \tau_j(l))\end{aligned} \tag{C.10}$$

$$\begin{aligned}\left[J_{\tau\tau}\right]_{ij} &= \mathrm{E}\left\{\left[\frac{\partial}{\partial\tau_i}\ln\Lambda(r)\right]\left[\frac{\partial}{\partial\tau_j}\ln\Lambda(r)\right]\right\} \\ &= \frac{1}{\sigma_\eta^2}\sum_{l=1}^{L} d_{i,m_l} d_{j,m_l}\beta_i(l)\beta_j(l)\bar{a}_i^d(lT_s - m_lT_b - \tau_i(l)) \\ &\qquad \cdot\bar{a}_j^d(lT_s - m_lT_b - \tau_j(l))\end{aligned} \tag{C.11}$$

where $\bar{a}_j^d(lT_s - m_lT_b - \tau_j(l))$ denotes the derivative of the filtered spreading code $\bar{a}_j(lT_s - m_lT_b - \tau_j(l))$ with respect to the delay τ_j.

By defining $\boldsymbol{B}(l) = \text{diag}[\boldsymbol{\beta}(l)]$ and $\bar{\boldsymbol{a}}_d(l)$ as in (6.86), it is easy to show that

$$
\begin{aligned}
\boldsymbol{J}_{\boldsymbol{\beta}\boldsymbol{\beta}} &= \frac{1}{\sigma_\eta^2}\sum_{l=1}^{L}\boldsymbol{\mathcal{D}}(l)\bar{\boldsymbol{a}}(l)\bar{\boldsymbol{a}}^T(l)\boldsymbol{\mathcal{D}}(l) \\
\boldsymbol{J}_{\boldsymbol{\beta}\boldsymbol{\tau}} &= \frac{1}{\sigma_\eta^2}\sum_{l=1}^{L}\boldsymbol{\mathcal{D}}(l)\bar{\boldsymbol{a}}(l)\bar{\boldsymbol{a}}_d^T(l)\boldsymbol{\mathcal{D}}(l)\boldsymbol{B}(l) \\
\boldsymbol{J}_{\boldsymbol{\tau}\boldsymbol{\tau}} &= \frac{1}{\sigma_\eta^2}\sum_{l=1}^{L}\boldsymbol{B}(l)\boldsymbol{\mathcal{D}}(l)\bar{\boldsymbol{a}}_d\bar{\boldsymbol{a}}_d^T\boldsymbol{\mathcal{D}}(l)\boldsymbol{B}(l) \ .
\end{aligned}
$$

The CRLB for the parameters $\boldsymbol{\beta}$ and $\boldsymbol{\tau}$ are found from the inverse of the Fischer information matrix which is given by

$$
\boldsymbol{J}^{-1} = \begin{bmatrix} \text{CRLB}(\boldsymbol{\beta}) & \boldsymbol{T} \\ \boldsymbol{T}^T & \text{CRLB}(\boldsymbol{\tau}) \end{bmatrix} \tag{C.12}
$$

where CRLB($\boldsymbol{\beta}$) and CRLB($\boldsymbol{\tau}$) are the Cramér-Rao bounds for the channel amplitudes and delays, respectively. The matrix $\boldsymbol{T}$ is of no interest to us since it does not influence the CRLBs in which we are interested. From (C.5) and using the standard result for the inverse of a partitioned matrix [71], the desired CRLBs are given by

$$
\begin{aligned}
\text{CRLB}(\boldsymbol{\beta}) &= \left[\boldsymbol{J}_{\boldsymbol{\beta}\boldsymbol{\beta}} - \boldsymbol{J}_{\boldsymbol{\beta}\boldsymbol{\tau}}\boldsymbol{J}_{\boldsymbol{\tau}\boldsymbol{\tau}}^{-1}\boldsymbol{J}_{\boldsymbol{\beta}\boldsymbol{\tau}}^T\right]^{-1} && \text{(C.13)} \\
\text{CRLB}(\boldsymbol{\tau}) &= \left[\boldsymbol{J}_{\boldsymbol{\tau}\boldsymbol{\tau}} - \boldsymbol{J}_{\boldsymbol{\beta}\boldsymbol{\tau}}^T\boldsymbol{J}_{\boldsymbol{\beta}\boldsymbol{\beta}}^{-1}\boldsymbol{J}_{\boldsymbol{\beta}\boldsymbol{\tau}}\right]^{-1} \ . && \text{(C.14)}
\end{aligned}
$$

Appendix D
Derivation of the GWRLS Algorithm

In this appendix, we develop the GWRLS algorithm used in Chapter 7. To facilitate development of the algorithm, we write (7.31) in matrix form as

$$\mathcal{J}(l) = \mathcal{E}^T(l)\mathcal{C}(l)\mathcal{E}(l) \tag{D.1}$$

where

$$\begin{aligned}
\mathcal{E}(l) &= \mathcal{R}_m(l) - \mathcal{H}(l)\boldsymbol{\theta}_s & \text{(D.2)}\\
&= \left[\mathbf{e}^T(1), \ldots, \mathbf{e}^T(l)\right]^T & \text{(D.3)}\\
&= [e_1(1), \ldots, e_N(1), \ldots, e_1(l), \ldots, e_N(l)]^T & \text{(D.4)}
\end{aligned}$$

and

$$\mathcal{C}(l) = \begin{bmatrix} \Lambda(1) & & \mathbf{0} \\ & \ddots & \\ \mathbf{0} & & \Lambda(l) \end{bmatrix} \tag{D.5}$$

where $\mathbf{e}(i)$ and $\Lambda(i)$ are defined in equations (7.27) and (7.32), respectively. Also,

$$\mathcal{R}_m(l) = \begin{bmatrix} \boldsymbol{r}_m(1) \\ \vdots \\ \boldsymbol{r}_m(l) \end{bmatrix} \quad \text{and} \quad \mathcal{H}(l) = \begin{bmatrix} \boldsymbol{H}_c(1) \\ \vdots \\ \boldsymbol{H}_c(l) \end{bmatrix} \tag{D.6}$$

where we note that $\mathcal{R}_m(l)$ implicitly depends on $\boldsymbol{\theta}_s$ through the measurements $\boldsymbol{r}_m(l)$. Hence, the cost function in equation (D.1), using all information up to time l, can be expressed as

$$\mathcal{J}(l) = [\mathcal{R}_m(l) - \mathcal{H}(l)\boldsymbol{\theta}_s]^T\mathcal{C}(l)[\mathcal{R}_m(l) - \mathcal{H}(l)\boldsymbol{\theta}_s] \ . \tag{D.7}$$

From this formulation, we can immediately express the LS estimate of $\boldsymbol{\theta}_s$ as

$$\hat{\boldsymbol{\theta}}_s = \left[\boldsymbol{\mathcal{H}}^T(l)\boldsymbol{\mathcal{C}}(l)\boldsymbol{\mathcal{H}}(l)\right]^{-1}\boldsymbol{\mathcal{H}}^T(l)\boldsymbol{\mathcal{C}}(l)\boldsymbol{\mathcal{R}}_m(l) \ . \quad \text{(D.8)}$$

Since we are interested in a recursive estimator, we partition the vectors and matrices as follows:

$$\boldsymbol{\mathcal{R}}_m(l+1) = \begin{bmatrix} \boldsymbol{\mathcal{R}}_m(l) \\ \boldsymbol{r}_m(l+1) \end{bmatrix} \qquad \boldsymbol{\mathcal{H}}(l+1) = \begin{bmatrix} \boldsymbol{\mathcal{H}}(l) \\ \boldsymbol{H}_c(l+1) \end{bmatrix} \quad \text{(D.9)}$$

$$\boldsymbol{\mathcal{C}}(l+1) = \begin{bmatrix} \lambda\boldsymbol{\mathcal{C}}(l) & \mathbf{0} \\ \mathbf{0} & \boldsymbol{W} \end{bmatrix} . \quad \text{(D.10)}$$

By defining the term in brackets in equation (D.8) as $\boldsymbol{P}(l)$, we can write

$$\begin{aligned} \boldsymbol{P}(l+1) &= \left[\boldsymbol{\mathcal{H}}^T(l+1)\boldsymbol{\mathcal{C}}(l+1)\boldsymbol{\mathcal{H}}(l+1)\right]^{-1} \\ &= \left[\lambda\boldsymbol{P}^{-1}(l) + \boldsymbol{H}_C^T(l+1)\boldsymbol{W}\boldsymbol{H}_C(l+1)\right]^{-1} \ . \quad \text{(D.11)} \end{aligned}$$

Using the Sherman-Morrison-Woodbury matrix inversion formula [58], we can express $\boldsymbol{P}(l+1)$ as

$$\begin{aligned} \boldsymbol{P}(l+1) &= \frac{1}{\lambda}\boldsymbol{P}(l) - \frac{1}{\lambda}\boldsymbol{P}(l)\boldsymbol{H}_c^T(l+1) \\ &\quad \cdot\left[\lambda\boldsymbol{W}^{-1} + \boldsymbol{H}_C(l+1)\boldsymbol{P}(l)\boldsymbol{H}_C^T(l+1)\right]^{-1} \\ &\quad \cdot\boldsymbol{H}_c(l+1)\boldsymbol{P}(l) \quad \text{(D.12)} \end{aligned}$$

from which we define

$$\boldsymbol{\Gamma}_{\text{rls}}(l+1) = \boldsymbol{P}(l)\boldsymbol{H}_C^T(l+1)\left[\lambda\boldsymbol{W}^{-1} + \boldsymbol{H}_C(l+1)\boldsymbol{P}(l)\boldsymbol{H}_C^T(l+1)\right]^{-1} . \quad \text{(D.13)}$$

Thus,

$$\boldsymbol{P}(l+1) = \frac{1}{\lambda}\left[\boldsymbol{I} - \boldsymbol{\Gamma}_{\text{rls}}(l+1)\boldsymbol{H}_C(l+1)\right]\boldsymbol{P}(l) \ . \quad \text{(D.14)}$$

Applying the result for $\boldsymbol{P}(l+1)$ to the estimate in equation (D.8) and using the partitioned vectors above, the estimate of $\boldsymbol{\theta}_s$ at time $l+1$ can be expressed as

$$\hat{\boldsymbol{\theta}}_s(l+1) = \hat{\boldsymbol{\theta}}_s(l) + \boldsymbol{\Gamma}_{\text{rls}}(l+1)\left[\boldsymbol{r}_m(l+1) - \boldsymbol{r}_m(l)|_{\hat{\boldsymbol{\theta}}_s(l)}\right] . \quad \text{(D.15)}$$

References

[1] J. Andersen, T. Rappaport, and S. Yoshida, "Propagation Measurements and Models for Wireless Communications Channels," *IEEE Communications Magazine*, Vol. 33, pp. 42–49, January 1995.

[2] B. Anderson and J. Moore, *Optimal Filtering*. Prentice-Hall, 1979.

[3] T. Aulin, "A Modified Model for the Fading Signal at a Mobile Radio Channel," *IEEE Trans. on Vehicular Technology*, Vol. 28, pp. 182–203, August 1979.

[4] M. Austin and G. Stüber, "Velocity Adaptive Handoff Algorithms for Microcellular Systems," *IEEE Trans. on Vehicular Technology*, Vol. 43, pp. 549–561, August 1994.

[5] Y. Bar-Shalom, *Estimation and Tracking: Principles, Techniques and Software*. Artech House, 1993.

[6] A. Barabell, "Improving the Resolution Performance of Eigenstructure-based Direction Finding Algorithms," in *International Conference on Acoustics Speech and Signal Processing*, pp. 336–339, 1983.

[7] S. Bensley and B. Aazhang, "Subspace-based Estimation in Multipath Channel Parameters for CDMA Communication Systems," in *IEEE Global Telecom. Conf., Comm. Theory Mini-Conf.*, pp. 154–157, 1994.

[8] S. Bensley and B. Aazhang, "Subspace-based Channel Estimation for Code Division Multiple Access Communication Systems," *IEEE Trans. on Communications*, Vol. 44, pp. 1009–1020, August 1996.

[9] J. Berg, R. Bownds, and F. Lotse, "Path Loss and Fading Models for Microcells at 900 MHz," in *IEEE Vehicular Technology Conference*, pp. 666–671, 1992.

[10] U. Bernhard, "Influence of Data Modulation and Doppler Effects on the Performance of a Delay Locked Loop," in *IEEE International Symposium on Spread Spectrum Techniques and Applications*, pp. 598–602, 1994.

[11] G. Beveridge and R. Schechter, *Optimization: Theory and Practice.* MacGraw-Hill Book Co., 1970.

[12] K. Blackard, M. Feuerstein, T. Rappaport, and S. Seidel, "Path Loss and Delay Spread Models as Functions of Antenna Height for Microcellular System Design," in *IEEE Vehicular Technology Conference*, pp. 333–337, 1992.

[13] W. Braun and U. Dersch, "A Physical Mobile Radio Channel," *IEEE Trans. on Vehicular Technology*, Vol. 40, pp. 472–482, May 1991.

[14] R. Bultitude and G. Bedal, "Propagation Characteristics on Microcellular Urban Mobile Radio Channels at 910 MHz," *IEEE Journal on Selected Areas in Communications*, Vol. 7, pp. 31–39, January 1989.

[15] J. Caffery and G. Stüber, "Effects of Multiple-Access interference on the Non-coherent Delay-Lock Loop," *Submitted for publication to IEEE Trans. Commun.*, 1997,1998.

[16] J. Caffery and G. Stüber, "Subscriber Location in CDMA Cellular Networks," *IEEE Trans. on Vehicular Technology*, Vol. 47, pp. 406–416, May 1998.

[17] J. Caffery, Jr. and G. Stüber, "Vehicle Location and Tracking for IVHS in CDMA Microcells," in *IEEE Personal Indoor Mobile Radio Conference*, pp. 1227–1231, 1994.

[18] J. Caffery, Jr. and G. Stüber, "Radio Location in Urban CDMA Microcells," in *IEEE Personal Indoor Mobile Radio Conference*, 1995.

[19] J. Caffery, Jr. and G. Stüber, "Overview of Radiolocation in CDMA Cellular Systems," *IEEE Communications Magazine*, Vol. 36, pp. 38–45, April 1998.

[20] G. Carter, ed., *Coherence and time delay estimation : an applied tutorial for research, development, test, and evaluation engineers.* IEEE Press, 1993.

[21] G. Carter, A. Nuttall, and P. Cable, "The Smoothed Coherence Transform," *Proc. of the IEEE*, Vol. 61, pp. 1479–1498, October 1973.

[22] CC Docket No. 94-102, "Revision of the Commissions Rules to Ensure Compatibility with Enhanced 911 Emergency Calling Systems, RM-8143," July 26 1996.

[23] C.-K. Chan and W.-H. Lam, "A Simplified Aperiodic Cross-Correlation Model for Direct-Sequence Spread-Spectrum Multiple-Access Communications Systems," in *IEEE International Conference on Communications*, pp. 1516–1520, 1994.

[24] Y. Chan, R. Hattin, and J. Plant, "The Least Squares Estimation of Time Delay and Its Use in Signal Detection," *IEEE Trans. on Acoustics, Speech, and Signal Processing*, Vol. ASSP-26, pp. 217–222, June 1978.

[25] Y. Chan and K. Ho, "An Efficient Closed-Form Localization Solution from Time Difference of Arrival Measurements," in *International Conference on Acoustics Speech and Signal Processing*, pp. II/393–396, 1994.

[26] K. Chawla and D. Sarwate, "Parallel Acquisition of PN Sequences in DS/SS Systems," in *IEEE International Conference on Communications*, pp. 28–32, 1992.

[27] P. Chestnut, "Emitter Location Accuracy Using TDOA and Differential Doppler," *IEEE Trans. on Aerospace and Electronic Systems*, Vol. AES-18, pp. 214–218, March 1982.

[28] J. Cisneros, D. Delley, and L. Greenbaum, "An Urban Positioning Approach Applying Differential Methods to Commercial FM Radio Emissions for Ground Mobile Users," in *Proc. ION Fifth Annual Meting*, pp. 83–92, June 1994.

[29] G. Corazza and V. Esposti, "Code Acquisition in DS-SSMA Systems in the Presence of Multiple Access Interference and Data Modulation," *European Trans. on Telecomm. and Related Tech.*, Vol. 5, pp. 27–37, February 1994.

[30] COST 207 TD(86)51-REV 3 (WG1), ed., *Proposal on channel transfer functions to be used in GSM late 1986*, September 1986.

[31] COST 231 TD(973)119-REV 2 (WG2), ed., *Urban Transmission Loss Models for Mobile Radio in the 900- and 1800-MHz Bands*, September 1991.

[32] D. Cox, "Wireless Personal Communications: A Perspective," in *The Mobile Communications Handbook*, (J. Gibson, ed.), ch. 15, pp. 209–241, CRC Pres, Inc., 1996.

[33] C. Drane and C. Rizos, *Positioning Systems in Intelligent Transportation Systems.* Artech House, 1998.

[34] A. Duel-Hallen, "A Family of Multiuser Decision-Feedback Detectors for Asynchronous Code-Division Multiple-Access Channels," *IEEE Trans. on Communications*, Vol. 43, pp. 421–434, Feb./Mar./Apr. 1995.

[35] L. Dumont, M. Fattouche, and G. Morrison, "Super-Resolution of Multipath Channels in a Spread Spectrum Location System," *Electron. Lett.*, Vol. 30, pp. 1583–1584, September 15 1994.

[36] P. Enge, "The Global Positioning System: Signals, Measurements, and Performance," *International Journal of Wireless Information Networks*, Vol. 1, pp. 83–105, 1994.

[37] J. Engel, "Effects of Multipath Transmission on the Measured Propagation Delay of an FM Signal," *IEEE Trans. on Vehicular Technology*, Vol. VT-18, pp. 44–52, May 1969.

[38] V. Erceg, S. Ghassemzadeh, M. Taylor, D. Li, and D. Schilling, "Urban/Suburban Out-of-Sight Propagation Modeling," *IEEE Communications Magazine*, Vol. 30, pp. 56–61, June 1992.

[39] J. V. Etten, "Navigation Systems: Fundamentals of Low- and Very-Low-Frequency Hyperbolic Techniques," *Electrical Communication*, Vol. 45, pp. 192–212, March 1970.

[40] B. Fang, "Simple Solutions for Hyperbolic and Related Position Fixes," *IEEE Trans. on Aerospace and Electronic Systems*, Vol. 26, pp. 784–753, September 1990.

[41] M. Feuerstein and T. Pratt, "A Local Area Position Location System," in *Fifth International Conference on Mobile Radio and Personal Communications*, pp. 79–83, 1989.

[42] W. Figel, N. Shepherd, and W. Trammell, "Vehicle Location by a Signal Attenuation Method," *IEEE Trans. on Vehicular Technology*, Vol. VT-18, pp. 105–110, November 1969.

[43] W. Foy, "Position-Location Solutions by Taylor-Series Estimation," *IEEE Trans. on Aerospace and Electronic Systems*, Vol. AES-12, pp. 187–193, March 1976.

[44] B. Friedlander, "A Passive Location Algorithm and its Accuracy Analysis," *IEEE Journal of Oceanic Engineering*, Vol. OE-12, pp. 234–244, January 1987.

[45] T. Fulghum and K. Molnar, "The Jakes Fading Model Incorporating Angular Spread for a Disk of Scatterers," in *IEEE Vehicular Technology Conference*, pp. 489–493, 1998.

[46] A. Fuxjaeger and R. Iltis, "Adaptive Parameter Estimation Using Parallel Kalman Filtering for Spread Spectrum Code and Doppler Tracking," *IEEE Trans. on Communications*, Vol. 42, pp. 2227–2230, June 1994.

[47] M. Gans, "A Power-Spectral Theory of Propagation in the Mobile-Radio Environment," *IEEE Trans. on Vehicular Technology*, Vol. VT-21, pp. 27–38, February 1972.

[48] W. Gardner, *Cyclostationarity in Communications and Signal Processing.* IEEE Press, 1994.

[49] W. Gardner and C. Chen, "Signal-Selective Time-Difference-of-Arrival Estimation for Passive Location of Man-Made Signal Sources in Highly Corruptive Environments, Part I: Theory and Method," *IEEE Trans. on Signal Processing*, Vol. 40, pp. 1168–1184, May 1992.

[50] W. Gardner and C. Chen, "Signal-Selective Time-Difference-of-Arrival Estimation for Passive Location of Man-Made Signal Sources in Highly Corruptive Environments, Part II: Algorithms and Performance," *IEEE Trans. on Signal Processing*, Vol. 40, pp. 1185–1197, May 1992.

[51] V. Garg, K. Smolik, and J. Wilkes, *Applications of CDMA in Wireless/Personal Communications.* Prentice Hall, 1997.

[52] P. Gething, *Radio Direction Finding and Superresolution.* Peter Peregrinus Ltd., 1991.

[53] A. Ghosh and R. Love, "Mobile Station Location in a DS-CDMA System," in *IEEE Vehicular Technology Conference*, pp. 254–258, 1998.

[54] A. Giordano, M. Chan, and H. Habal, "A Novel Location-Based Service and Architecture," in *IEEE Personal Indoor Mobile Radio Conference*, pp. 853–857, 1995.

[55] P. Goud, A. Sesay, and M. Fattouche, "A Spread Spectrum Radiolocation Technique and Its Application to Cellular Radio," in *IEEE Pacific Rim Conference on Communications, Computers and Signal Processing*, pp. 661–664, 1991.

[56] E. Green, "Path Loss and Signal Variability Analysis for Microcells," in *Fifth International Conference on Mobile Radio and Personal Communications*, pp. 38–42, 1989.

[57] L. Greenstein, N. Amitay, T.-S. Chu, L. Cimini, Jr., G. Foschini, M. Gans, A. Rustako, Jr., R. Valenzuela, and G. Vannucci, "Microcells in Personal Communications Systems," *IEEE Communications Magazine*, Vol. 30, pp. 76–88, December 1992.

[58] M. Grewal and A. Andrews, *Kalman Filtering: Theory and Practice.* Prentice-Hall: Englewood Cliffs, NJ, 1993.

[59] M. Gudmundson, "Analysis of Handover Algorithms," in *IEEE Vehicular Technology Conference*, pp. 537–541, 1991.

[60] W. Hahn and S. Tretter, "Optimum Processing for Delay-Vector Estimation in Passive Signal Arrays," *IEEE Trans. on Information Theory*, Vol. IT-19, pp. 608–614, September 1973.

[61] P. Harley, "Short Distance Attenuation Measurements at 900 MHz and 1.8 GHz Using Low Antenna Heights for Microcells," *IEEE Journal on Selected Areas in Communications*, Vol. 7, pp. 5–13, January 1989.

[62] H. Hashemi, "Pulse Ranging Radiolocation Technique and Its Application to Channel Assignment in Digital Cellular Radio," in *IEEE Vehicular Technology Conference*, pp. 675–680, 1991.

[63] M. Hata and T. Nagatsu, "Mobile Location Using Signal Strength Measurements in a Cellular System," *IEEE Trans. on Vehicular Technology*, Vol. VT-29, pp. 245–251, May 1980.

[64] S. Haykin, *Adaptive Filter Theory.* Prentice Hall, 1991.

[65] E. Hepsaydir and W. Yates, "Performance Analysis of Mobile Positioning Using Existing CDMA Network," in *IEEE Int. Symposium on Spread Spectrum Techniques and Applications*, pp. 190–192, 1994.

[66] K. Ho and Y. Chan, "Solution and Performance Analysis of Geolocation by TDOA," *IEEE Trans. on Aerospace and Electronic Systems*, Vol. 29, pp. 1311–1322, October 1993.

[67] J. Holmes, *Coherent Spread Spectrum Systems.* John Wiley & Sons, Inc., 1982.

[68] J. Holtzman, "DS/CDMA Successive Interference Cancellation," in *IEEE Int. Symposium on Spread Spectrum Techniques and Applications*, pp. 69–78, 1994.

[69] J. Holtzmann, S. Nanda, and D. Goodman, "CDMA Power Control for Wireless Networks," in *Third Generation of Wireless Information Networks*, pp. 299–311, Kluwer Academic Publishers, 1992.

[70] S. Hong, S. Yoon, H. Lee, and J. Ahn, "Performance Analysis of Non-coherent Delay-Locked Loop in Multiple Access Interference," *IEICE Trans. on Commun.*, Vol. E78, pp. 935–941, June 1995.

[71] R. Horn and C. Johnson, *Matrix Analysis.* Cambridge University Press, 1985.

[72] B. Ibrahim and A. Aghvami, "Direct Sequence Spread Spectrum Matched Filter Acquisition in Frequency-Selective Rayleigh Fading Channels," *IEEE Journal on Selected Areas in Communications*, Vol. 12, pp. 885–890, June 1994.

[73] B. Ibrahim and A. Aghvami, "A PN Code Tracking System for Direct Sequence Spread Spectrum Operation in a Frequency Selective Fading Environment," in *IEEE Global Telecommunications Conference*, pp. 374–378, 1994.

[74] *IEEE Communications Magazine*, "Issue on Wireless Geolocation Systems and Services," Vol. 36, April 1998.

[75] *IEEE Trans. Acoustics, Speech and Signal Processing*, "Special Issue on Time-Delay Estimation," Vol. ASSP-29, June 1981.

[76] R. Iltis, "Joint Estimation of PN Code Delay and Multipath Using the Extended Kalman Filter," *IEEE Trans. on Communications*, Vol. 38, pp. 1677–1685, October 1990.

[77] R. Iltis and A. Fuxjaeger, "A Digital DS Spread-Spectrum Receiver with Joint Channel and Doppler Shift Estimation," *IEEE Trans. on Communications*, Vol. 39, pp. 1255–1267, August 1991.

[78] R. Iltis and L. Mailaender, "An Adaptive Multiuser Detector with Joint Amplitude and Delay Estimation," *IEEE Journal on Selected Areas in Communications*, Vol. 12, pp. 774–785, June 1994.

[79] ITS America, "Strategic Plan for Intelligent Vehicle Highway Systems in the United States," 1992.

[80] W. C. Jakes, *Microwave Mobile Communications.* IEEE Press, 1994.

[81] J. Joutsensalo, "Algorithms for Delay Estimation and Tracking in CDMA," in *IEEE Vehicular Technology Conference*, pp. 366–370, 1997.

[82] S. Julier and J. Uhlmann, "A New Extension of the Kalman Filter to Nonlinear Systems," in *International Symposium on Aerospace/Defense Sensing, Simulation and Controls*, 1997.

[83] S. Julier, J. Uhlmann, and H. Durrant-Whyte, "A New Approach for Filtering Nonlinear Systems," in *American Control Conference*, pp. 1628–32, 1995.

[84] R. Jurgen, "The Electronic Motorist," *IEEE Spectrum*, Vol. 32, pp. 37–48, March 1995.

[85] C. Knapp and G. Carter, "The Generalized Correlation Method for Estimation of Time Delay," *IEEE Trans. on Acoustics, Speech, and Signal Processing*, Vol. ASSP-24, pp. 320–327, August 1976.

[86] K. Kosbar and J. Zaninovich, "Periodic PN Sequence Delay Estimation Using Phase Spectrum Data," in *IEEE Global Telecommunications Conference*, pp. 1665–1669, 1993.

[87] K. Krizman, T. Biedka, and T. Rappaport, "Wireless Position Location: Fundamentals, Implementation Strategies, and Sources of Error," in *IEEE Vehicular Technology Conference*, pp. 919–923, 1997.

[88] R. Kumaresan and D. Tufts, "Estimating the angles of arrival of multiple plane waves," *IEEE Trans. on Aerospace and Electronic Systems*, Vol. AES-19, pp. 134–139, January 1983.

[89] M. Latva-aho and J. Lilleberg, "Delay Trackers for Multiuser CDMA Receivers," in *IEEE International Conference on Universal Personal Communications*, pp. 326–330, 1996.

[90] H. Lee, "Accuracy Limitations of Hyperbolic Multilateration Systems," *IEEE Trans. on Aerospace and Electronic Systems*, Vol. AES-11, pp. 16–29, January 1975.

[91] W. Lee, *Mobile Communications Engineering*. McGraw Hill, 1982.

[92] W. Lee, "Estimate of Local Average Power of a Mobile Radio Signal," *IEEE Trans. on Vehicular Technology*, Vol. VT-34, pp. 22–27, February 1985.

[93] W. Lee, "Elements of Cellular Mobile Radio Systems," *IEEE Trans. on Vehicular Technology*, Vol. VT-35, pp. 48–56, May 1986.

[94] W. Lee, "Overview of Cellular CDMA," *IEEE Trans. on Vehicular Technology*, Vol. 40, pp. 291–301, May 1991.

[95] W. Lee, "Power Control in CDMA (cellular radio)," in *IEEE Vehicular Technology Conference*, pp. 77–80, 1991.

[96] W. Lee, "Lee's Model," in *IEEE Vehicular Technology Conference*, pp. 343–348, 1992.

[97] J. Liberti, *Analysis of CDMA Cellular Radio Systems Employing Adaptive Antennas*. Ph. D. thesis, Virginia Tech, September 1995.

[98] J. Liberti and T. Rappaport, "Accurate Techniques to Evaluate CDMA Bit Error Rates in Multipath Channels with Imperfect Power Control," in *IEEE Global Telecommunications Conference*, pp. 33–37, 1995.

[99] C. Lo, K. Chen, and W. Sheen, "Noncoherent DLL and TDL PN Code Tracking Loops in Rayleigh Fading Channels," in *IEEE Personal Indoor Mobile Radio Conference*, pp. 338–342, 1994.

[100] M. Lötter and P. van Rooyen, "A Probability Density Function for the Angle of Arrival of Signals in Cellular CDMA/SDMA Systems," in *IEEE Int. Symposium on Spread Spectrum Techniques and Applications*, pp. 227–231, 1998.

[101] R. Lupas and S. Verdú, "Linear Multiuser Detectors for Synchronous Code-Division Multiple-Access Channels," *IEEE Trans. on Information Theory*, Vol. IT-35, pp. 123–126, January 1989.

[102] R. Lupas and S. Verdú, "Near-Far Resistance of Multiuser Detectors in Asynchronous Channels," *IEEE Trans. on Communications*, Vol. 39, pp. 496–508, April 1990.

[103] M. Madfors, K. Wallstedt, S. Magnusson, H. Olofsson, P.-. Backman, and S. Engström, "High Capacity with Limited Spectrum in Cellular Systems," *IEEE Communications Magazine*, Vol. 35, pp. 38–45, August 1997.

[104] U. Madhow and M. Pursley, "Acquisition in Direct-Sequence Spread-Spectrum Communication Networks: an Asymptotic Analysis," *IEEE Trans. on Information Theory*, Vol. 39, pp. 903–912, May 1993.

[105] T. Manickam, R. Vacaro, and D. Tufts, "Least-Squares Algorithm for Multipath Time-Delay Estimation," *IEEE Trans. on Signal Processing*, Vol. 42, pp. 3229–3233, November 1994.

[106] N. Marchand, "Error Distributions of Best Estimate of Position from Multiple Time Difference Hyperbolic Networks," *IEEE Transactions on Aerospace and Navigational Electronics*, Vol. 11, pp. 96–100, June 1964.

[107] M. Marsan and G. Hess, "Shadow Variability in an Urban Land Mobile Radio Environment," *Electron. Lett.*, Vol. 26, pp. 646–648, May 1990.

[108] P. Massatt and K. Rudnick, "Geometric Formulas for Dilution of Precision Calculations," *Journal of the Institute of Navigation*, Vol. 37, pp. 379–391, Winter 1991.

[109] P. Maybeck, *Stochastic Models, Estimation, and Control.* Vol. 2, Academic Press, 1979.

[110] L. Melin, M. Rönnlund, and r. Angbratt, "Radio Wave Propagation: A comparison Between 900 and 1800 MHz," in *IEEE Vehicular Technology Conference*, pp. 250–252, 1993.

[111] H. Messer and Y. Bar-Ness, "Closed-Loop Least Mean Square Time-Delay Estimator," *IEEE Trans. on Acoustics, Speech, and Signal Processing*, Vol. ASSP-35, pp. 413–424, April 1987.

[112] H. Meyr, "Delay-Lock Tracking of Stochastic Signals," *IEEE Trans. on Communications*, Vol. COM-24, pp. 331–339, March 1976.

[113] P. Mogensen, P. Eggers, C. Jensen, and J. Andersen, "Urban Area Radio Propagation Measurements at 955 and 1845 Mhz for Small and Micro Cells," in *IEEE Global Telecommunications Conference*, pp. 1297–1302, 1991.

[114] T. Moon, R. Short, and C. Rushforth, "Average Acquisition Time for SSMA Channels," in *MILCOM '91*, pp. 1042–1045, 1991.

[115] G. Morley and W. Grover, "Improved Location Estimation with Pulse-Ranging in Presence of Shadowing and Multipath Excess-Delay Effects," *Electron. Lett.*, Vol. 31, pp. 1609–1610, Aug. 21 1995.

[116] M. Napier and V. Ashkenazi, "Modern Navigation and Positioning Techniques," *Navigation*, Vol. 40, pp. 183–193, 1987.

[117] Y. Okumura, E. Ohmori, T. Kawano, and K. Fukuda, "Field Strength and its variability in VHF and UHF Land Mobile Service," *Rev. Electr. Commun. Lab.*, Vol. 16, pp. 825–873, 1968.

[118] G. Ott, "Vehicle Location in Cellular Mobile Radio Systems," *IEEE Trans. on Vehicular Technology*, Vol. VT-26, pp. 43–46, February 1977.

[119] J. Padgett, C. Günther, and T. Hattori, "Overview of Wireless Personal Communications," *IEEE Communications Magazine*, Vol. 33, pp. 28–41, January 1995.

[120] M. Pallas and G. Jourdain, "Active High Resolution Time Delay Estimation for Large BT Signals," *IEEE Trans. on Signal Processing*, pp. 781–787, April 1991.

[121] J. Parsons, *The Mobile Radio Propagation Channel.* Halsted Press, 1992.

[122] I. Paton, E. Crompton, J. Gardner, and J. Noras, "Terminal Self-Location in Mobile Radio Systems," in *Sixth International Conference on Mobile Radio and Personal Communications*, pp. 203–207, 1991.

[123] M. Pätzold and F. Laue, "Statistical Properties of Jakes' Fading Channel Simulator," in *IEEE Vehicular Technology Conference*, pp. 712–718, 1998.

[124] A. Paulraj, R. Roy, and T. Kailath, "Estimation of Signal Parameters via Rotational Invariance Techniques – ESPRIT," in *Proceedings of Asilomar Conf. on Circuits and Systems*, pp. 83–89, 1986.

[125] Paulraj, et. al., *Subspace Methods for Directions-of-Arrival Estimation*, ch. 16. Elsevier Science Publishers, 1993.

[126] R. Pickholtz, L. Milstein, and D. Schilling, "Spread Spectrum for Mobile Communications," *IEEE Trans. on Vehicular Technology*, Vol. 40, pp. 313–321, May 1991.

[127] R. Pickholtz, D. Schilling, and L. Milstein, "Theory of Spread-Spectrum Communications—A Tutorial," *IEEE Trans. on Communications*, Vol. COM-30, pp. 855–884, May 1982.

[128] J. Pierce, "Omega," *IEEE Trans. on Aerospace and Electronic Systems*, Vol. AES-1, pp. 206–215, December 1965.

[129] A. Piersol, "Time Delay Estimation Using Phase Data," *IEEE Trans. on Acoustics, Speech, and Signal Processing*, Vol. ASSP-29, pp. 471–477, June 1981.

[130] A. Polydros and C. Weber, "Analysis and Optimization of Correlative Code-Tracking Loops in Spread-Spectrum Systems," *IEEE Trans. on Communications*, Vol. COM-33, pp. 30–43, January 1985.

[131] C. Powell, "The Decca Navigator System for Ship and Aircraft Use," *Institution of Electrical Engineers, Proceedings*, Vol. 105, pp. 225–34, 277–283, March 1958.

[132] B. Quine, J. Uhlmann, and H. Durrant-Whyte, "Implicit Jacobians for Linearised State Estimation in Nonlinear Systems," in *American Control Conference*, pp. 1645–46, 1995.

[133] A. Radovic, "An Iterative Near-Far Resistant Algorithm for Joint Parameter Estimation in Asynchronous CDMA Systems," in *Proc. 5th International Symposium Personal, Indoor, Mobile Radio Conference*, pp. 199–203, 1994.

[134] T. Rappaport, *Wireless Communications: Principles and Practice*. Prentice Hall, 1996.

[135] T. Rappaport, J. Reed, and B. Woerner, "Position Location Using Wireless Communications on the Highways of the Future," *IEEE Communications Magazine*, Vol. 34, pp. 33–41, October 1996.

[136] Y. Rasshcheplyayev and V. Shcherbachev, "Estimation of Coordinates in Range-Difference Radar Systems on the Basis of Projective Transformations of Observation Equations," *Journal of Communications Technology and Electronics*, Vol. 39, pp. 1627–1636, 1995.

[137] J. Reed, K. Krizman, B. Woerner, and T. Rappaport, "An Overview of the Challenges and Progress in Meeting the E-911 Requirement for Location Service," *IEEE Communications Magazine*, Vol. 36, pp. 30–37, April 1998.

[138] D. Reudink, "Properties of Mobile Radio Propagation above 400 MHz," *IEEE Trans. on Vehicular Technology*, Vol. VT-23, pp. 143–159, November 1974.

[139] S. Rhee, "Vehicle Location in Angular Sectors Based on Signal Strength," *IEEE Trans. on Vehicular Technology*, Vol. VT-27, pp. 244–258, November 1978.

[140] R. Rick and L. Milstein, "Noncoherent Parallel Acquisition in CDMA Spread Spectrum Systems," in *IEEE International Conference on Communications*, pp. 1422–1426, 1994.

[141] S. Riter and J. McCoy, "Vehicle Location—An Overview," *IEEE Trans. on Vehicular Technology*, Vol. VT-26, pp. 7–11, February 1977.

[142] P. Roth, "Effective Measurements using Digital Signal Analysis," *IEEE Spectrum*, Vol. 8, pp. 62–70, April 1971.

[143] A. Rustako, Jr., N. Amitay, G. Owens, and R. Roman, "Radio Propagation at Microwave Frequencies for Line-of-Sight Microcellular Mobile Personal Communications," *IEEE Trans. on Vehicular Technology*, Vol. 40, pp. 203–120, February 1991.

[144] S. Sakagami, *et. al.*, "Vehicle Position Estimates by Multibeam Antennas in Multipath Environments," *IEEE Trans. on Vehicular Technology*, Vol. 41, pp. 63–67, February 1992.

[145] H. Saarnisaari, "TLS-ESPRIT in a Time Delay Estimation," in *IEEE Vehicular Technology Conference*, pp. 1619–1623, 1997.

[146] S. Schell, Calabretta, W. Gardner, and B. Agee, "Cyclic MUSIC Algorithms for Signal Selective DOA Estimation," in *International Conference on Acoustics Speech and Signal Processing*, pp. 2278–2281, 1988.

[147] R. Schmidt, "A New Approach to Geometry of Range Difference Location," *IEEE Trans. on Aerospace and Electronic Systems*, Vol. AES-8, pp. 821–835, November 1972.

[148] R. Schmidt, *A Signal Subspace Approach to Multiple Emitter Location and Spectral Estimation.* Ph. D. thesis, Dept. of Elec. Eng., Stanford Univ., November 1981.

[149] R. Scholtz, "The Spread Spectrum Concept," *IEEE Trans. on Communications*, Vol. COMM-25, pp. 748–755, August 1977.

[150] T. Shan, M. Wax, and T. Kailath, "On Spatial Smoothing for Estimation of Coherent Signals," *IEEE Trans. on Acoustics, Speech, and Signal Processing*, Vol. ASSP-33, pp. 802–811, August 1985.

[151] W. Sheen and G. Stüber, "Effects of Multipath Fading on Delay-Locked Loops for Spread Spectrum Systems," *IEEE Trans. on Communications*, Vol. 42, pp. 1947–1956, February/March/April 1994.

[152] W. Sheen and G. Stüber, "A New Tracking Loop for Direct Sequence Spread Spectrum Systems on Frequency-Selective Fading Channels," *IEEE Trans. on Communications*, Vol. 43, pp. 3063–3072, December 1995.

[153] Z.-L. Shi, P. Driessen, and W. Du, "PN Code Acquisition for Asynchronous CDMA Communications Based on Interference Cancellation," in *IEEE Pacific Rim Conference*, pp. 778–781, 1993.

[154] M. Silventoinen and T. Rantalainen, "Mobile Station Emergency Locating in GSM," in *IEEE Int. Conference on Personal Wireless Communications*, pp. 232–238, 1996.

[155] Simon, Omura, Scholtz, and Levitt, *Spread Spectrum Communications.* Vol. III, Computer Science Press, 1985.

[156] B. Sklar, "Rayleigh Fading Channels in Mobile Digital Communication Systems Part I: Characterization," *IEEE Communications Magazine*, Vol. 35, pp. 90–100, July 1997.

[157] J. Smith and J. Abel, "Closed-Form Least Squares Source Location Estimation from Range-Difference Measurements," *IEEE Trans. on Acoustics, Speech, and Signal Processing*, Vol. ASSP-35, pp. 1661–1669, December 1987.

[158] J. Smith and J. Abel, "The Spherical Interpolation Method of Source Localization," *IEEE Journal of Oceanic Engineering*, Vol. OE-12, pp. 246–252, January 1987.

[159] W. Smith, Jr., "Passive Location of Mobile Cellular Telephone Terminals," in *IEEE International Carnahan Conference on Security Technology*, pp. 221–225, 1991.

[160] H. So and P. Cheng, "Target Localisation in Presence of Multipaths," *Electron. Lett.*, Vol. 29, pp. 293–294, February 4 1993.

[161] H.-L. Song, "Automatic Vehicle Location in Cellular Communications Systems," *IEEE Trans. on Vehicular Technology*, Vol. 43, pp. 902–908, November 1994.

[162] E. Sourour and S. Gupta, "Direct-Sequence Spread-Spectrum Parallel Acquisition in a Fading Mobile Channel," *IEEE Trans. on Communications*, Vol. 38, pp. 992–998, July 1990.

[163] J. Spilker, "Delay-Lock Tracking of binary Signals," *IEEE Trans. on Space Electronics and Telemetry*, Vol. SET-9, pp. 1–8, March 1963.

[164] J. Spilker, *Digital Communications by Satellite.* Prentice-Hall, 1977.

[165] J. Spilker, Jr. and D. Magill, "The Delay-Lock Discriminator—An Optimum Tracking Device," *Proceedings IRE*, Vol. 49, pp. 1403–1416, September 1961.

[166] H. Staras and S. Honickman, "The Accuracy of Vehicle Location by Trilateration in a Dense Urban Environment," *IEEE Trans. on Vehicular Technology*, Vol. VT-21, pp. 38–43, February 1972.

[167] H. Stark and J. Woods, *Probability, Random Processes, and Estimation Theory for Engineers.* Prentice-Hall, Inc., 1994.

[168] State of New Jersey, "Report on the New Jersey Wireless Enhanced 911 System Trial: The First 100 Days," June 16 1997.

[169] R. Steele, *Mobile Radio Communications.* Pentech Press, 1994.

[170] L. Stilp, "Time Difference of Arrival Technology for Locating Narrowband Cellular Signals," in *Proceedings of the SPIE*, pp. 134–144, 1995.

[171] L. Stilp, "Carrier and End-User Applications for Wireless Location Systems," in *Proceedings of the SPIE*, pp. 119–126, 1996.

[172] E. Ström, S. Parkvall, S. Miller, and B. Ottersten, "Sensitivity Analysis of Near-Far Resistant DS-CDMA Receivers to Propagation Delay Estimation Errors," in *IEEE Vehicular Technology Conference*, pp. 757–761, 1994.

[173] E. Ström, S. Parkvall, S. Miller, and B. Ottersten, "DS-CDMA Synchronization in Time-Varying Fading Channels," *IEEE Journal on Selected Areas in Communications*, pp. 1636–1642, October 1996.

[174] E. Ström, S. Parkvall, S. Miller, and B. Ottersten, "Propagation Delay Estimation in Asynchronous Direct-Sequence Code-Division Multiple Access Systems," *IEEE Trans. on Communications*, Vol. 44, pp. 84–93, January 1996.

[175] G. Stüber, *Principles of Mobile Communication.* Kluwer Academic Publishers, 1996.

[176] S.-L. Su, N.-Y. Yen, and S.-C. Hsieh, "Performance Analysis of Digital Delay Lock Loop in the Presence of Doppler Shift," *IEEE Trans. on Communications*, Vol. 44, pp. 668–674, June 1996.

[177] M. Takeuchi, A. Kajiwara, and M. Nakagawa, "A Delay Lock Loop for Mobile Communications in the presence of Multipath Fading," *IEICE Trans. on Commun.*, Vol. E76-B, pp. 1039–1045, August 1993.

[178] The Strategis Group, "Wireless Location Services, 1997," 1997.

[179] *Mobile Station-Base Station Compatibility Standard for Dual-Mode Wideband Spread Spectrum Digital Cellular System.* TIA/EIA IS-95, PN-3422 1994.

[180] D. Torrieri, "Statistical Theory of Passive Location Systems," *IEEE Trans. on Aerospace and Electronic Systems*, Vol. AES-20, pp. 183–197, March 1984.

[181] G. Turin, W. Jewell, and T. Johnston, "Simulation of Urban Vehicle-Monitoring Systems," *IEEE Trans. on Vehicular Technology*, Vol. VT-21, pp. 9–16, February 1972.

[182] A. Turkmani, J. Parsons, F. Ju, and D. Lewis, "Microcellular Radio Measurements at 900, 1500 and 1800 MHz," in *Fifth International Conference on Mobile Radio and Personal Communications*, pp. 65–68, 1989.

[183] A. van Dierendonck, P. Fenton, and T. Ford, "Theory and Performance of Narrow Correlator Spacing in a GPS Receiver," *Journal of the Institute of Navigation*, Vol. 39, pp. 265–283, Fall 1992.

[184] D. van Nee, "Multipath Effects on GPS Code Phase Measurements," *Journal of the Institute of Navigation*, Vol. 39, pp. 177–190, Summer 1992.

[185] D. van Nee, "The Multipath Estimating Delay Lock Loop: approaching Theoretical Accuracy Limits," in *IEEE Position, Location and Navigation Symposium*, pp. 246–251, 1994.

[186] H. van Trees, *Detection, Estimation, and Linear Modulation Theory.* John Wiley and sons, 1968.

[187] M. Varanasi and B. Aazhang, "Multistage Detection in Asynchronous Code-Division Multiple-Access Communications," *IEEE Trans. on Communications*, Vol. COM-38, pp. 509–519, April 1990.

[188] S. Verdú, *Optimum Multiuser Signal Detection.* Ph. D. thesis, Dept. of Elec. Comp. Eng., Univ. of Illinois, Urban-Champaign, August 1984.

[189] S. Verdú, "Minimum Probability of Error for Asynchronous Gaussian Multiple-Access Channels," *IEEE Trans. on Information Theory*, Vol. IT-32, pp. 85–96, January 1986.

[190] G. D. Vitantonio and F. D. Ferrari, "On the Application of Near-Far Resistant Techniques to DS-CDMA Systems," in *IEEE Personal Indoor Mobile Radio Conference*, pp. 70–74, 1994.

[191] A. Viterbi, *CDMA: Principles of Spread Spectrum Communication.* Addison-Wesley, 1995.

[192] R. Ward, "Acquisition of Pseudonoise Signals by Sequential Estimation," *IEEE Trans. on Communications*, Vol. COM-13, pp. 475–483, December 1965.

[193] B. Washburn, "Wireless Applications for Intelligent Transportation Systems," *Telecommunications*, Vol. 30, pp. 40–44, July 1996.

[194] M. Wax, "The Joint Estimation of Differential Delay, Doppler, and Phase," *IEEE Trans. on Information Theory*, Vol. IT-28, pp. 817–820, September 1982.

[195] A. Wilde, "Extended Tracking Range Delay-Locked Loop," in *IEEE International Conference on Communications*, pp. 1051–1054, 1995.

[196] T. Wisløff and S. Andresen, "Positioning and Traffic Information by Cellular Radio," in *IEEE Vehicle Navigation and Information Systems Conference*, pp. 287–290, 1993.

[197] M. Wylie and J. Holtzmann, "The Non-Line of Sight Problem in Mobile Location Estimation," in *IEEE International Conference on Universal Personal Communications*, pp. 827–831, 1996.

[198] Xia, et. al., "Radio Propagation Measurements and Modeling for Line-of-Sight Microcellular Systems," in *IEEE Vehicular Technology Conference*, pp. 349–354, 1992.

[199] Z. Xie, C. Rushforth, and R. Short, "Multiuser Signal Detection Using Sequential Decoding," *IEEE Trans. on Communications*, Vol. 38, pp. 578–583, May 1990.

[200] K. Yao, "Error Probability of Asynchronous Spread Spectrum Multiple Access Communication Systems," *IEEE Trans. on Communications*, Vol. COM-25, pp. 803–809, August 1977.

[201] Y. Yoon, R. Kohno, and H. Imai, "A Spread Spectrum Multi-Access System with a Cascade of Co-Channel Interference Cancelers for Multipath Fading Channels," in *IEEE Int. Symposium on Spread Spectrum Techniques and Applications*, pp. 87–90, 1992.

[202] G. Yost and S. Panchapakesan, "Automatic Location Identification Using a Hybrid Technique," in *IEEE Vehicular Technology Conference*, pp. 276–267, 1998.

[203] R. Yost and R. Boyd, "A Modified PN Code Tracking Loop: Its Performance Analysis and Comparative Evaluation," *IEEE Trans. on Communications*, Vol. COM-30, pp. 1027–1036, May 1982.

[204] Y. Zhao, *Vehicle Location and Navigation Systems*. Artech House, 1997.

[205] W. Zhaocheng, W. Jing, Y. Zhixing, and Y. Yan, "Synchronization Consideration in Multiuser CDMA Environment," in *Int. Conf. on Communication Tech.*, pp. 103–106, 1996.

[206] F. Zheng and S. Barton, "Performance Analysis of Near-Far Resistant CDMA Detectors – Influence of System Parameter Estimation

Errors," in *IEEE International Conference on Communications*, pp. 520–524, 1994.

[207] R. Ziemer and R. Peterson, *Digital Communications and Spread Spectrum Systems.* Macmillan Publishing Co., 1985.

[208] I. Ziskind and M. Wax, "Maximum Likelihood Localization of Multiple Sources by Alternating Projection," *IEEE Trans. on Acoustics, Speech, and Signal Processing*, Vol. 36, pp. 1553–1560, October 1988.

About the Author

James Caffery, Jr., was born in Alton, IL, in May 1970. He received the B.S.E.E degree (summa cum laude) from Bradley University, Peoria, IL, in 1992, and the M.S. and Ph.D. degrees from the Georgia Institute of Technology in 1993 and 1998, respectively.

From 1992 to 1998, he worked as a Graduate Research Assistant studying wireless location in CDMA cellular communications systems for his doctoral research. During the summer of 1996, he worked with GTE Mobilnet, Atlanta, GA, where he was a part of the Technology Research group. In 1999, he joined the faculty of the Department of Electrical and Computer Engineering and Computer Science at the University of Cincinnati as an Assistant Professor. His research interests include CDMA communications systems, spread spectrum, wireless location, parameter estimation, and multiuser communication theory. He is a member of the IEEE, Eta Kappa Nu, and Phi Kappa Phi.

Index

MIX
Papier | Fördert
gute Waldnutzung
FSC® C083411

Kolibri 360 GmbH
Ferdinand-Jühlke-Straße 7
99095 Erfurt, Deutschland
produktsicherheit@kolibri360.de